How to Lie with Maps

How to Lie with Maps

THIRD EDITION

Mark Monmonier

The University of Chicago Press

CHICAGO AND LONDON

The University of Chicago Press, Chicago 60637
The University of Chicago Press, Ltd., London
© 1991, 1996, 2018 by The University of Chicago
Published 2018
Printed in the United States of America

27 26 25 24 23   6

ISBN-13: 978-0-226-43592-3 (paper)
ISBN-13: 978-0-226-43608-1 (e-book)
DOI: https://doi.org/10.7208/chicago/[9780226436081].001.0001

LIBRARY OF CONGRESS CATALOGING-IN-PUBLICATION DATA

Names: Monmonier, Mark, author.
Title: How to lie with maps / Mark Monmonier.
Description: Third edition. | Chicago ; London : The University of
Chicago Press, 2018. | Includes bibliographical references and index.
Identifiers: LCCN 2017039382| ISBN 9780226435923 (pbk. : alk.
paper) | ISBN 9780226436081 (e-book)
Subjects: LCSH: Cartography. | Deception.
Classification: LCC G108.7 .M66 2018 | DDC 526—dc23
LC record available at https://lccn.loc.gov/2017039382

♾ This paper meets the requirements of ANSI/NISO Z39.48-1992
(Permanence of Paper).

For Marge and Jo, again

Contents

Preface to the Third Edition

More than twenty years have passed between the second and third editions of this book. For all that has changed in the world of mapping during that time—including, above all, radical technological changes in how maps are made, displayed, and used—the basic principles set forth in the first two editions are as relevant as ever: maps offer believable representations of reality, people tend to trust them even though oftentimes one map cannot tell the whole story, and healthy skepticism is essential because map authors who don't understand or otherwise ignore cartographic principles can commit misleading blunders. Consequently maps can indeed lie in multiple ways, inadvertently if not deliberately.

When I embarked on this new edition, I was determined to preserve the essential character of the book while updating the text (and the graphics) for a new generation of readers. As one of the geographers who reviewed my revision proposal pointed out, most of his students have never used a paper map for regular navigation, a perspective this edition has taken into account. While most of the issues discussed in the previous editions are still relevant (with some adjustments) to digital as well as paper maps, overhead imagery and online mapping have also introduced new issues requiring coverage. I decided to replace the outdated chapter "Multimedia, Experiential Maps, and Graphic Scripts" with three new chapters on image maps, prohibitive cartography, and "fast maps," the latter category encompassing animated maps, interactive maps, web maps, and clever static maps disseminated electronically on the internet—maps that go viral are, indisputably, fast maps. The result is a modest increase in the book's length but an important improvement in its scope. Most of the images added to the enhanced color insert are also related to these new chapters.

Although many readers first encountered *How to Lie with Maps* as an assigned text in a college course on cartography, map reading, basic geography, or geographic information systems (GIS), I continue to envision my target audience as the intelligent lay reader who is curious about maps, a population that has only grown over time with the proliferation of maps in all media. And as graphics software has increasingly allowed map users to become mapmakers in their own right, I hope this book will guide them to make more informed cartographic choices.

Mark Monmonier
DeWitt, New York

Introduction

Not only is it easy to lie with maps, it's essential. To portray meaningful relationships for a complex, three-dimensional world on a flat sheet of paper or a screen, a map must distort reality. As a scale model, the map must use symbols that almost always are proportionally much bigger or thicker than the features they represent. To avoid hiding critical information in a fog of detail, the map must offer a selective, incomplete view of reality. There's no escape from the cartographic paradox: to present a useful and truthful picture, an accurate map must tell white lies.

Because most map users willingly tolerate white lies on maps, it's not difficult for maps to also tell more serious lies. Map users have traditionally been a trusting lot: they understand the need to distort geometry and suppress features, and they believe the cartographer really does know where to draw the line, figuratively as well as literally. As with many things beyond their full understanding, they generally entrust mapmaking to a priesthood of technically competent designers and geographers working for government agencies and commercial firms. Yet cartographers are not licensed, and many mapmakers competent in commercial art or in the use of graphics software have never studied cartography. Even map users who know that such software is widely available and who see maps in an increasing range of media seldom question these authorities, and they often fail to appreciate the map's power as a tool of deliberate falsification or subtle propaganda.

Because anyone with the right software and an internet connection can now make and publish maps, mapmakers can also easily lie to themselves and others—and be unaware of it. Before the electronic age, folk cartography consisted largely of hand-drawn maps giving directions. The direction giver had full control over pencil and paper and usually had no difficulty transferring routes, landmarks, and other relevant recollections from mind to map, but the result was clearly amateurish. Technology allows people without cartographic savvy to create modern-day folk maps with the crisp type, uniform symbols, and verisimilitude of maps from the cartographic priesthood. Yet software developers have also made it easy for the lay cartographer to select an inappropriate projection or a misleading set of symbols. Because of advances in graphics software and online mapping, inadvertent yet serious cartographic lies can appear respectable and accurate.

The potential for cartographic mischief extends well beyond the deliberate manipulations used by some cartographer-propagandists and the electronic blunders made by the cartographically ignorant. If any single caveat can alert map users to their unhealthy but widespread naïveté, it is that *a single map is but one of an indefinitely large number of maps that might be produced for the same situation or from the same data.* The italics reflect an academic lifetime of browbeating undergraduates with this obvious but readily ignored warning. How easy it is to forget—and how revealing to recall—that map authors can experiment freely with features, measurements, area of coverage, and symbols and can pick the map that best presents their case or supports their unconscious bias. Map users must be aware that cartographic license is enormously broad.

The purpose of this book is to promote a healthy skepticism about maps, not to foster either cynicism or deliberate dishonesty. In showing how to lie with maps, I want to make readers aware that maps, like speeches and paintings, are authored collections of information and are also subject to distortions arising from ignorance, greed, ideological blindness, or malice.

Examining the misuses of maps—both paper and digital—also provides an interesting introduction to the nature of maps and their range of appropriate uses. The four chapters that follow this one address general cartographic principles that apply to all different types of maps. Chapter 2 considers the map's main elements—scale, projection, and symbolization—as potential sources of distortion. Chapter 3 further explores the effects of scale by examining the various white lies cartographers justify as necessary generalization, and chapter 4 looks at common blunders resulting from the mapmaker's ignorance or oversight. Chapter 5 looks at how a careless or Machiavellian choice of colors can confuse or mislead the map viewer.

The rest of the chapters treat specific types of maps and how they can be manipulated. Chapter 6 treats the seductive use of symbols in advertising maps, and chapter 7 explores exaggeration and suppression in maps prepared for development plans and environmental-impact statements. Chapters 8 and 9 examine distorted maps used by governments as political propaganda and as "disinformation" for military opponents. Government mapping is also a central concern in chapter 10, which investigates the effects of national culture and bureaucratic inertia—and, increasingly, commercial interests—on detailed topographic maps. Chapter 11 addresses distortion and self-deception in maps made from census and survey data and other quantitative information. Chapter 12 examines the specific challenges posed by image maps that are based on satellite technology and other measurements, chapter 13 acknowledges the emergence of prohibitive mapping as a pervasive and potentially threatening cartographic genre, and chapter 14 addresses the diverse types of dynamic maps and the distinctive advantages and constraints of online maps. Chapter 15 concludes by noting how maps can have dual and sometimes conflicting roles and by recommending a skeptical assessment of a map author's motives.

In an era of increasing skepticism about the nature of knowledge, a book about what it means to lie with maps is more useful than ever. For all the interest in verbal lies, nefarious as well as white, and in how words can be manipulated, education in the use of maps and other visuals is spotty and limited, and many otherwise-educated people are graphically and cartographically illiterate. Maps, like numbers, are often arcane images accorded undue respect and credibility. This book's principal goal is to dispel this cartographic mystique and promote a more informed use of maps based on an understanding and appreciation of their flexibility as a medium of communication.

As technology continues to lower the barriers dividing map users from mapmakers, this book's insights can be especially useful for those who might more effectively use maps in their work or as citizens fighting environmental deterioration or social ills. The informed skeptic becomes a perceptive map author, better able to describe locational characters and explain geographic relationships and better equipped to recognize and counter the self-serving arguments of biased or dishonest mapmakers.

Elements of the Map

Maps have three basic attributes: scale, projection, and symbolization. Each element is a source of distortion. As a group, they describe the essence of the map's possibilities and limitations. No one can use maps or make maps safely and effectively without understanding map scales, map projections, and map symbols.

Scale

Most maps are smaller than the reality they represent, and map scales tell us how much smaller. A map can state its scale in three ways: as a ratio, as a short sentence, or as a simple graph. Figure 2.1 shows some typical statements of map scale.

Ratio scales relate one unit of distance on the map to a specific distance on the ground. The units must be the same, so that a ratio of 1:10,000 means that a 1-inch line on the map represents a 10,000-inch stretch of road—or that 1 centimeter represents 10,000 centimeters or 1 foot stands for 10,000 feet. As long as they are the same, the units don't matter and need not be stated; the ratio scale is a dimensionless number. By convention, the part of the ratio to the left of the colon is always 1.

Some maps state the ratio scale as a fraction, but both forms have the same meaning. Whether the mapmaker uses 1:24,000 or 1/24,000 is solely a matter of style.

Fractional statements help the user compare map scales. A scale of 1/10,000 (or 1:10,000) is larger than a scale of 1/250,000 (or 1:250,000) because 1/10,000 is a larger fraction than 1/250,000.

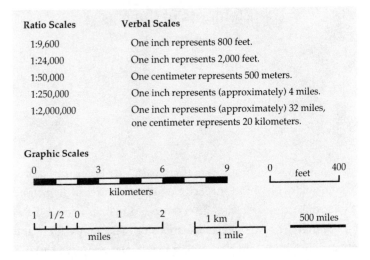

Figure 2.1. Types of map scales.

Recall that small fractions have big denominators and big fractions have small denominators, or that half (1/2) of a pie is more than a quarter (1/4) of the pie. In general, "large-scale" maps have scales of 1:24,000 or larger, whereas "small-scale" maps have scales of 1:500,000 or smaller. But these distinctions are relative: in a city-planning office where the smallest map scale is 1:50,000, "small-scale" might refer to maps at 1:24,000 or smaller and "large-scale" to maps at 1:4,800 or larger.

Large-scale maps tend to be more detailed than small-scale maps. Consider two maps, one at 1:10,000 and the other at 1:10,000,000. A 1-inch line at 1:10,000 represents 10,000 inches, which is 833 1/3 feet, or roughly 0.16 miles. At this scale a square measuring 1 inch on each side represents an area of 0.025 square miles, or roughly 16 acres. In contrast, at 1:10,000,000 the 1-inch line on the map represents almost 158 miles, and the square inch would represent an area slightly over 24,900 square miles, or nearly 16 million acres. In this example the square inch on the large-scale map could show features on the ground in far greater detail than the square inch on the small-scale map. Both maps would have to suppress some details, but the designer of the 1:10,000,000-scale map must be far more selective than the cartographer producing the 1:10,000-scale map. In the sense that all maps tell white lies about the planet, small-scale maps have a smaller capacity for truth than large-scale maps.

Verbal statements such as "one inch represents one mile" relate units convenient for measuring distances on the map to units commonly used for estimating and thinking about distances on the ground. For most users this simple sentence is more meaningful than the corresponding ratio scale of 1:63,360, or its close approximation, 1:62,500. British map users used to identify various map series with adjective phrases such as "inch to the mile" or "four miles to the inch" (a close approximation for 1:250,000).

Sometimes a mapmaker might say "equals" instead of "represents." Although technically absurd, "equals" in these cases might more kindly be considered a shorthand for "is the equivalent of." Yet the skeptic rightly warns of cartographic seduction, for "one inch equals one mile" not only robs the user of a subtle reminder that the map is merely a symbolic model but also falsely suggests that the mapped image is reality. As later chapters show, this delusion can be dangerous.

Metric units make verbal scales less necessary. Persons familiar with centimeters and kilometers have little need for sentences to tell them that at 1:100,000 one centimeter represents one kilometer, or that at 1:25,000 four centimeters represent one kilometer. In Europe, where metric units are standard, round-number map scales of 1:10,000, 1:25,000, 1:50,000, and 1:100,000 are common. In the United States, where the metric system's most prominent inroads have been in the liquor and drug businesses, large-scale maps typically represent reality at scales of 1:9,600 ("one inch represents eight hundred feet"), 1:24,000 ("one inch represents two thousand feet"), and 1:62,500 ("one inch represents [slightly less than] one mile").

Graphic scales are not only the most helpful means of communicating map scale but also the safest. An alternative to blind trust in the user's sense of distance and skill in mental arithmetic, the simple bar scale typically portrays a series of conveniently rounded distances appropriate to the map's function and the area covered. Graphic scales are particularly safe for maps that might be reduced or enlarged for publication, or by users. For example, a 5-inch-wide map labeled "1:50,000" would have a scale less than 1:80,000 if reduced to fit a newspaper column or a mobile-device screen that is 3 inches wide, whereas a scale bar representing a half mile would shrink along with the map's other symbols and distances. Ratio and verbal scales are useless on digital maps, since screens and thus the map scales vary widely and unpredictably.

Web-based maps and similar interactive applications occasion a fourth type of map scale: the zoom slider that moves up or down to indicate relative distance above the surface, or the interactive plus and minus buttons that produce the same effect. Zooming out yields a broader geographic scope with a smaller scale and less detail, and zooming in provides a narrower view with greater detail.

Map Projections

Map projections, which transform the curved, three-dimensional surface of the planet into a flat, two-dimensional plane, can greatly distort map scale. Although the globe can be a true scale model of the earth, with a constant scale at all points and in all directions, the flat map stretches some distances and shortens others, so that scale varies from point to point. Indeed, the bar scale sometimes included in the lower right of an online map can be blatantly misleading when the user zooms out to show an entire continent. Moreover, scale at a point tends to vary with direction as well.

The world-map projection in figure 2.2 illustrates the often severe scale differences found on maps portraying large areas. In this instance map scale is constant along the equator and the meridians, which are shown as straight lines perpendicular to the equator and running from the North Pole to the South Pole. (If

Figure 2.2. Equatorial cylindrical projection with true meridians.

the terms *parallel, meridian, latitude,* and *longitude* seem puzzling, the quick review of basic world-geographic concepts found in the appendix might be helpful.) Because the meridians have the same scale as the equator, each meridian (if we assume the earth is a *perfect* sphere) is half the length of the equator. Because scale is constant along the meridians, the map preserves the even spacing of parallels separated by 30° of latitude. But on this map all parallels are the same length, even though on the earth or a globe parallels decrease in length from the equator to the poles. Moreover, the map projection has stretched the poles from points with no length to lines as long as the equator. North-south scale is constant, but east-west scale increases to twice the north-south scale at 60° N and 60° S and to infinity at the poles.

Ratio scales commonly describe a world map's capacity for detail. But the scale is strictly valid for just a few lines on the map—in the case of figure 2.2, it is valid only for the equator and the meridians. Most world maps don't warn that using the scale ratio to convert distances between map symbols to distances between real places almost always yields an erroneous result. Figure 2.2, for instance, would greatly inflate the distance between Chicago and Stockholm, which are far apart and both well north of the equator. Cartographers wisely avoid decorating world maps with graphic

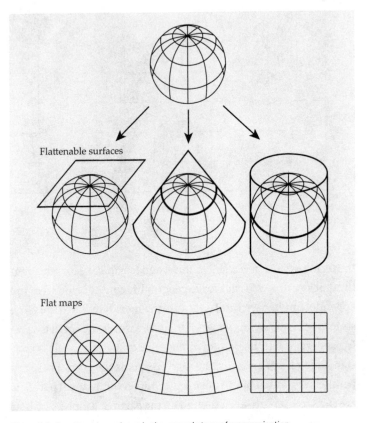

Figure 2.3. Developable surfaces in the second stage of map projection.

scales, which might encourage this type of abuse. In contrast, scale distortion of distance usually is negligible on large-scale maps, where the area covered is comparatively small.

Figure 2.3 helps explain the meaning and limitations of ratio scales in world maps by treating map projection as a two-stage process. Stage one shrinks the earth to a globe, for which the ratio scale is valid everywhere and in all directions. Stage two projects symbols from the globe onto a flattenable surface, such as a plane, a cone, or a cylinder, which is attached to the globe at a point or at one or two *standard lines*. On flat maps, the scale usually is constant only along these standard lines. In figure 2.2, a type of cylindrical

Secant cylindrical

Tangent cylindrical

Figure 2.4. Secant (above) and tangent (below) cylindrical projections.

projection called a *plane chart*, the equator is a standard line and the meridians show true scale as well.

In general, scale distortion increases with distance from the standard line. The common *developable surfaces*—plane, cone, and cylinder—allow the mapmaker to minimize distortion by centering the projection in or near the region featured on the map. World maps commonly use a cylindrical projection, centered on the equator. Figure 2.4 shows that a *secant* cylindrical projection, which cuts through the globe, yields two standard lines, whereas a *tangent* cylindrical projection, which merely touches the globe, has only one. Average distortion is less for a secant projection because the

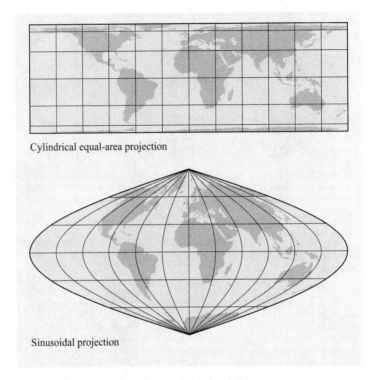

Cylindrical equal-area projection

Sinusoidal projection

Figure 2.5. Two varieties of equal-area cylindrical projection.

average place is closer to one of the two standard lines. Conic projections are well suited to large mid-latitude areas, such as North America, Europe, and Russia, and secant conic projections offer less average distortion than tangent conic projections. *Azimuthal* projections, which use the plane as their developable surface, are used most commonly for maps of polar regions.

For each developable surface, the mapmaker can choose from a variety of projections, each with a unique pattern of distortion. Some projections, called *equivalent* or *equal-area*, allow the mapmaker to preserve areal relationships. Thus if South America is eight times larger than Greenland on the globe, it will also be eight times larger on an equal-area projection. Figure 2.5 shows two ways to reduce the areal distortion of the plane chart (fig. 2.2). The

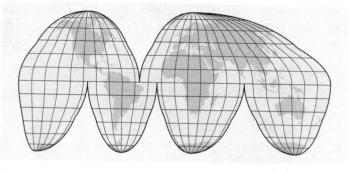

Figure 2.6. Goode's Homolosine Equal-Area projection.

cylindrical equal-area projection at the top compensates for the severe poleward exaggeration by reducing the separation of the parallels as distance from the equator increases. In contrast, the sinusoidal projection below maintains true scale along the equator, all other parallels, and the central meridian and at the same time pulls the meridians inward, toward the poles, compensating for the areal exaggeration that would otherwise occur. Distortion is least pronounced in a cross-shaped zone along the equator and the central meridian and most severe between these axes toward the edge of the projection. Despite the highly distorted shapes in these "corners," the areas of continents, countries, and belts between adjoining parallels are in correct proportion.

Reduced distortion around the central meridian suggests that a sinusoidal projection "centered" on a meridian through, say, Kansas might yield a decent equal-area representation of North America, whereas a sinusoidal projection with a straight-line central meridian passing between Warsaw and Moscow would afford a suitable companion view of the Eurasian land mass. In the early 1920s, University of Chicago geography professor J. Paul Goode extended this notion of a zoned world map and devised the composite projection in figure 2.6. Goode's Interrupted Homolosine Equal-Area projection has six lobes, which join along the equator. To avoid severe pinching of the meridians toward the poles, Goode divided each lobe into two zones at about 40°—an equatorial zone based on the sinusoidal projection and a poleward zone in which the equal-area Mollweide projection portrays high-latitude areas with less east-west compression. Goode's projection mollifies the trade-off of more distorted shapes for true relative areas by giving up continuous oceans for less severely distorted landmasses. If interrupted over the land to minimize distortion of the oceans, Goode's projection can be equally adept at serving studies of fisheries and other marine elements.

No flat map can match the globe in preserving areas, angles, gross shapes, distances, and directions, and any map projection is a compromised solution. Yet Goode's projection is a particularly worthy compromise when the mapmaker uses dot symbols to portray the worldwide density pattern of population, hogs, wheat, or other dryland variables. On a dot-distribution map with one dot representing five hundred thousand swine, for example, the spacing of these dots represents relative density. Important hog-producing regions, such as the American Midwest and southeastern China, have many closely spaced dots, whereas hog-poor regions, such as India and Australia, have few. But a projection that distorts area might show contrasting densities for two regions of equal size on the globe and with similar levels of hog production; if both regions have forty dots representing twenty million swine, the region occupying 2 square centimeters of the map would have a greater spacing between dots and appear less intensively involved in raising pigs than the region occupying only 1 square centimeter. Projections that are not equal-area encourage such spurious inferences. Equivalence is also important when the map user might compare the sizes of countries or the areas covered by various map categories.

As equal-area projections preserve areas, *conformal* projections preserve local angles. That is, on a conformal projection the angle between any two intersecting lines will be the same on both a globe and a flat map. By compressing three-dimensional physical features onto a two-dimensional surface, a conformal projection can noticeably distort the shapes of long features, but within a

Mercator projection   Gnomonic projection

Figure 2.7. Straight lines on an equatorially centered Mercator projection (left) are rhumb lines, which show constant geographic direction, whereas straight lines on a gnomonic projection (right) are great circles, which show the shortest route between two points.

small neighborhood of the point of intersection, scale will be the same in all directions and shape will be correct. Thus tiny circles on the globe remain tiny circles on a conformal map. As with all projections, though, scale still varies from place to place, and tiny circles identical in size on the globe can vary markedly in size on a conformal projection covering a large region. Although all projections distort the shapes of continents and other large territories, in general a conformal projection offers a less distorted picture of gross shape than a projection that is not conformal.

Perhaps the most striking trade-off in map projection is between conformality and areal equivalence. Although some projections distort both angles and areas, no projection can be both conformal and equivalent. Not only are these properties mutually exclusive, but in parts of the map well removed from the standard line(s) conformal maps severely exaggerate area and equal-area maps severely distort shape.

Two conformal projections useful in navigation illustrate how badly a map can distort area. The Mercator projection, on the left side of figure 2.7, renders Greenland as large as South America, whereas a globe would show Greenland as only about one-eighth

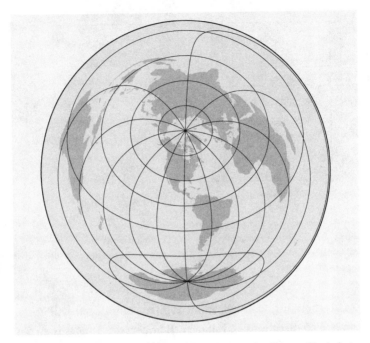

Figure 2.8. Oblique azimuthal equidistant projection centered on Chicago, Illinois, just east of the meridian at 90° W.

as large. North-south scale increases so sharply toward the poles that the poles themselves lie at infinity and never appear on an equatorially centered Mercator map. The right side of figure 2.7 reveals an even more severe distortion of area on the gnomonic projection, which cannot portray even half the globe.

Why, then, are these projections used at all? Although presenting two of the worst possible perspectives for general-purpose base maps and wall maps, these maps are of enormous value to a navigator with a straightedge. On the Mercator map, for instance, a straight line is a *rhumb line* or *loxodrome*, which shows an easily followed route of constant bearing. A navigator at A can draw a straight line to B, measure with a protractor the angle between this rhumb line and the meridian, and use this bearing and a corrected compass to sail or fly from A to B. On the gnomonic map, in contrast, a straight line represents a *great circle* and shows the shortest course from A to B. An efficient navigator would identify a few intermediate points on this great-circle route, transfer these course-adjustment points from the gnomonic map to the Mercator map, mark a chain of rhumb lines between successive intermediate points, measure each rhumb line's bearing, and proceed from A to B along a compromise course of easily followed segments that collectively approximate a shortest-distance route.

Map projections distort five geographic relationships: areas, angles, gross shapes, distances, and directions. Although some projections preserve local angles but not areas, others preserve areas but not local angles. All distort large shapes noticeably (but some distort continental shapes more than others), and all distort at least some distances and some directions. Yet, as the Mercator and gnomonic maps demonstrate, the mapmaker often can tailor the projection to serve a specific need. For instance, the oblique azimuthal *equidistant* projection in figure 2.8 shows true distance and directional relationships for shortest-distance great-circle routes converging on Chicago, Illinois. Although highly useful for someone concerned with relative proximity to Chicago, this projection is of no use for distance comparisons not involving Chicago. Moreover, its poor portrayal of the shapes and relative areas of continents, especially when extended to a full-world map, limits its value as a general-purpose reference map. With an interactive graphics system and good mapping software, of course, map users can become their own highly versatile mapmakers and tailor projections to many unique needs. For instance, an azimuthal equidistant map centered on North Korea might better inform a discussion of possible missile attacks by a rogue nation than concentric circles drawn thoughtlessly on an off-the-shelf world map.

Among the more highly tailored map projections are *cartograms*, which portray such relative measures as travel time, transport cost, and population size. Although a more conventional map might

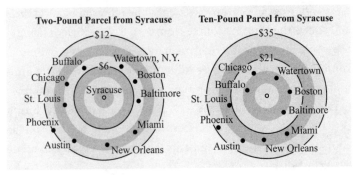

Figure 2.9. Distance cartograms showing relative spaces based on parcel-post rates from or to Syracuse, New York.

address these with tailored symbols and a standard projection, the geometry and layout of the cartogram make a strong visual statement of distance or area relationships. The *distance cartograms* in figure 2.9, for example, provide a dramatic comparison of two postal rates, which define different transport-cost spaces for their focal point, Syracuse, New York. Note that the rate for a 2-pound parcel mailed to Watertown, New York, is a little more than half the rate from Syracuse to Phoenix, Arizona, whereas the corresponding rates for a 10-pound parcel more nearly reflect Watertown's relative proximity (only 70 miles north of Syracuse). These schematic maps omit boundaries and other traditional frame-of-reference features, which are less relevant here than the names of the destinations shown.

Coastlines and some national boundaries are more useful in figure 2.10, an *area cartogram*, which even includes a pseudogrid to create the visual impression of "the world on a torus." This projection is a *demographic base map*, on which the relative sizes of areal units represent population, not land area. Note that the map portrays India as vastly larger than Canada because the Indian population is more than thirty-five times larger than the Canadian population, even though Canada's area of 3.8 million square miles is markedly larger than India's 1.2 million square miles. The cartogram has merged some countries with smaller populations,

Figure 2.10. "World on a Torus" demographic base map is an area cartogram based on the populations of major countries.

demonstrating the mapmaker's political insensitivity in sacrificing nationalism for clarity. Yet traditionalist cartographers who scorn cartograms as foolish, inaccurate cartoons ignore the power of map distortions to address a wide array of communicational and analytical needs.

Map Symbols

Graphic symbols complement map scale and projection by making visible the features, places, and other locational information represented on the map. By describing and differentiating features and places, map symbols serve as a graphic code for storing and retrieving data in a two-dimensional geographic framework. This code can be simple and straightforward, as on a route map drawn to show a new neighbor how to find the local elementary school: a few simple lines, labels, and Xs representing selected streets and landmarks should do. Labels such as "Elm St." and "Fire Dept." tie the map to reality and make a key or legend unnecessary.

When the purpose of the map is specific and straightforward, selection of map features also serves to suppress unimportant information. But maps mass-produced by government mapping agencies and commercial map publishers must address a wide variety of questions, and the maps' symbols must tell the user what's relevant and what isn't. Without the mapmaker present to explain unfamiliar details, these maps need a symbolic code based on an understanding of graphic logic and the limitations of visual perception. A haphazard choice of symbols, adequate for the labels and little pictures of way-finding maps and other folk cartography, can fail miserably on general-purpose maps rich in information.

Some maps, such as geologic maps and weather charts, have complex but standardized symbologies that organize an enormous amount of data meaningful only to those who understand the field and its cartographic conventions. Although as arcane to most people as a foreign language or higher mathematics, these maps also benefit from symbols designed according to principles of logic and communication.

Appreciating the logic of map symbols begins with understanding the three geometric categories of map symbols and the six visual variables shown in figure 2.11. Symbols on flat maps are either point symbols, line symbols, or area symbols. Road maps and most other general-purpose maps use combinations of all three: point symbols to mark the locations of landmarks and villages, line symbols to show the lengths and shapes of rivers and roads, and area symbols to depict the form and size of state parks and major cities. By contrast, *statistical maps*, which portray numerical data, commonly rely on a single type of symbol, such as dots that each denote ten thousand people or graytones representing election results by county.

Maps need contrasting symbols to portray geographic differences. As figure 2.11 illustrates, map symbols can differ in size, shape, graytone value, texture, orientation, and hue—that is, color differences, as between blue, green, and red (see chapter 5). Each of these six visual variables excels in portraying one kind of geo-

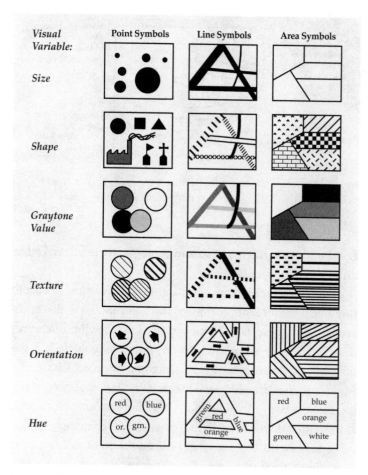

Visual Variable:	Point Symbols	Line Symbols	Area Symbols
Size			
Shape			
Graytone Value			
Texture			
Orientation			
Hue	red blue / or. grn.	green red blue orange	red blue / orange / green white

Figure 2.11. The six principal visual variables.

graphic difference. Shape, texture, and hue are effective in showing qualitative differences, as between types of land use or dominant religions. For quantitative differences, size is more suited to showing variation in amount or count, such as the number of television viewers by market area, whereas graytone value is preferred for portraying differences in rate or intensity, such as the proportion of the viewing audience watching the seventh game of the World Series. Symbols varying in orientation are useful mostly for rep-

Closely spaced contour lines indicate relatively steep slope.

Widely separated contour lines indicate relatively gentle slope.

Contour lines represent constant elevation.

Contour lines run perpendicular to slope. Arrows show downhill direction.

Figure 2.12. Elevation contours use two visual variables: spacing (texture) portrays steepness, and contour orientation is perpendicular to the direction of slope.

resenting winds, migration streams, troop movements, and other directional occurrences.

Some visual variables are unsuitable for small point symbols and thin line symbols that provide insufficient contrast with a background. Hue, for instance, is more effective at showing differences in kind for area symbols than for tiny point symbols, such as the dots on a dot-distribution map. Graytone value, which usually works well in portraying percentages and rates for area symbols, is visually less effective with point and line symbols, which tend to be thinner than area symbols. Point symbols commonly rely on shape to show differences in kind and on size to show differences in amount. Line symbols usually use hue or texture to distinguish rivers from railways and town boundaries from dirt roads. Size is useful in representing magnitude for links in a network: a thick line readily suggests greater capacity or heavier traffic than a thin line implies. Area symbols usually are large enough to reveal differences in hue, graytone, and pattern, but a detail inset, with a larger scale, might be needed to show very small yet important areal units.

Some symbols combine two visual variables. For example, the elevation contours on a topographic map involve both orientation and spacing, an element of pattern. As figure 2.12 demonstrates, a contour line's direction indicates the local direction of slope because the land slopes downward perpendicular to the trend of the contour line. And the spacing of the contour lines shows the relative tilt of the land because close contours mark steep slopes and separated contours indicate gentle slopes. Similarly, the spread of dots on a dot-distribution map may show the relative sizes of hog-producing regions, whereas the spacing or clustering of these dots reveals the relative intensity and geographic concentration of production.

A poor match between the data and the visual variable can frustrate or confuse the map user. Among the worst offenders are novice mapmakers seduced by the brilliant colors of electronic displays and color printers into using reds, blues, greens, yellows, and oranges to portray quantitative differences. Contrasting hues, however visually dramatic, are not an appropriate substitute for a logical series of easily ordered graytones. Except among physicists and professional "colorists," who understand the relation between hue and wavelength of light, map users cannot easily and consistently organize colors into an ordered sequence. And those with imperfect color vision might not even distinguish reds from greens. Yet most map users can readily sort five or six graytones evenly spaced between light gray and black; decoding is simple when darker means more and lighter means less. A legend might make a bad map useful, but it can't make it efficient.

Area symbols are not the only ones useful for portraying numerical data for states, counties, and other areal units. If the map must emphasize magnitudes such as the number of inhabitants rather than intensities such as the number of persons per square mile, point symbols varying in size are more appropriate than area symbols varying in graytone. The two areal-unit maps in figure 2.13 illustrate the different graphic strategies required for portraying population size and population density. The map on the left uses *graduated point symbols* positioned near the center of each area; the size of the point symbol represents population size. At its right a *choropleth map* uses graytone symbols that fill the areal units; the relative darkness of the symbol shows the concentration of population on the land.

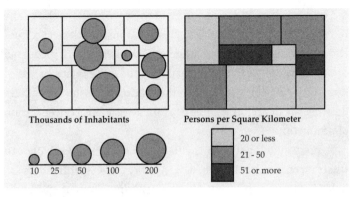

Figure 2.13. Graduated point symbols (left) and graytone area symbols (right) offer straightforward portrayals of population size and population density.

Because the visual variables match the measures portrayed, these maps are straightforward and revealing. At the left, big point symbols represent large populations, which occur in both large and small areas, and small point symbols represent small populations. On the choropleth map to the right, a dark symbol indicates many people occupying a relatively small area, whereas a light symbol represents either relatively few people in a small area or many people spread rather thinly across a large area.

Figure 2.14 illustrates the danger of an inappropriate match between measurement and symbol. Both maps portray population size, but the choropleth map at the right is misleading because its area symbols suggest intensity, not magnitude. Note, for instance, that the dark graytone representing a large county with a large but relatively sparsely distributed population also represents a small county with an equally large but much more densely concentrated population. In contrast, the map at the left provides not only a more direct symbolic representation of population size but a clearer picture of area boundaries and area size. The map user should beware of spurious choropleth maps based on magnitude yet suggesting density or concentration.

Form and color make some map symbols easy to decode. Pic-

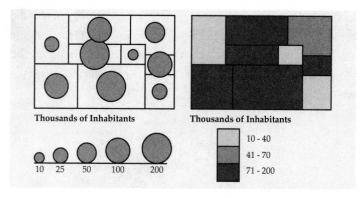

Figure 2.14. The map with graduated point symbols (left) using symbol size to portray magnitude demonstrates an appropriate choice of visual variable. The map with gray-tone area symbols (right) is ill-suited to portray magnitude.

torial point symbols effectively exploit familiar forms, as when little tents represent campgrounds and tiny buildings with crosses on top indicate churches. Alphabetic symbols also use form to promote decoding, as with common abbreviations ("PO" for post office), place-names ("Baltimore"), and labels describing the type of feature ("Union Pacific Railroad"). Color conventions allow map symbols to exploit idealized associations of lakes and streams with a bright, nonmurky blue and wooded areas with a wholesome, springlike green. Weather maps take advantage of perceptions of red as warm and blue as cold. Similarly, dashed lines might connote uncertainty in the location of a geologic fault line, while mildly transparent area symbols avoid graphic conflict with useful line symbols otherwise omitted.

Color codes often rely more on convention than on perception, as with land-use maps, where red commonly represents retail sales and blue stands for manufacturing. Physical-political reference maps found in atlases and on schoolroom walls reinforce the convention of *hypsometric tints*, a series of color-coded elevation symbols ranging from greens to yellows to browns. Although highly useful for those who know the code, elevation tints invite misinterpretation among the unwary. The greens used to represent lowlands, for instance, might suggest lush vegetation, whereas the browns representing highlands can connote barren land—despite the many lowland deserts and highland forests throughout the world. Like map projections, map symbols can lead naive users to wrong conclusions.

CHAPTER 3

Map Generalization: Little White Lies and Lots of Them

A good map tells a multitude of little white lies; it suppresses truth to help the user see what needs to be seen. Reality is three-dimensional, rich in detail, and far too factual to allow a complete yet uncluttered two-dimensional graphic scale model. Indeed, a map that did not generalize would be useless. But the value of a map depends on how well its generalized geometry and generalized content reflect a chosen aspect of reality.

Geometry

Clarity demands geometric generalization because map symbols usually occupy proportionately more space on the map than the features they represent occupy on the ground. For instance, a line 1/50 inch wide representing a road on a 1:100,000-scale map is the graphic equivalent of a corridor 167 feet wide. If a road's actual right-of-way was only 40 feet wide, say, a 1/50-inch-wide line symbol would claim excess territory at scales smaller than 1:24,000. At 1:100,000, this road symbol would crowd out sidewalks, houses, lesser roads, and other features. And at still smaller scales more important features might eliminate the road itself. These more important features could include national, state, or county boundaries, which have no width whatsoever on the ground.

Point, line, and area symbols require different kinds of generalization. For instance, cartographers recognize the five fundamental processes of geometric line generalization described in figure 3.1. First, of course, is the *selection* of complete features for the map. Selection is a positive term that implies the suppres-

27

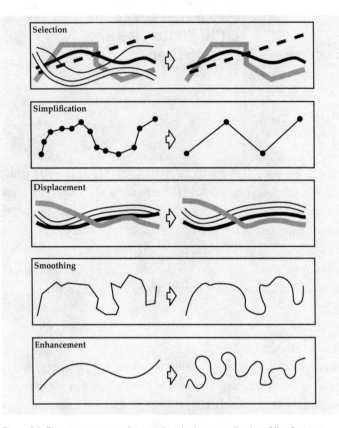

Figure 3.1. Elementary geometric operations in the generalization of line features.

sion, or nonselection, of most features. Ideally the map author approaches selection with goals to be satisfied by a well-chosen subset of all possible features that might be mapped and by map symbols chosen to distinguish unlike features and provide a sense of graphic hierarchy. Features selected to support the specific theme for the map usually require more prominent symbols than background features, chosen to provide a geographic frame of reference. Selecting background details that are effective in relating new information on the map to the viewer's geographic savvy and existing "mental map" often requires more insight and attention than selecting the map's main features. In the holistic process of planning a map, feature selection is the prime link between generalization and overall design.

The four remaining generalization processes in figure 3.1 alter the appearance and spatial position of linear map features that are represented by a series of points stored electronically as a list of two-dimensional (X, Y) coordinates. Although the growing use of software to generalize maps has led to the isolation of these four generalization operations, traditional cartographers perform essentially the same operations manually, at a graphics workstation or with pen and ink, but typically do so with less structure, less formal awareness, and less consistency than a computer algorithm. *Simplification*, which reduces detail and angularity by eliminating points from the list, is particularly useful if excessive detail was "captured" in developing a cartographic data file or if data developed for display at one scale are to be displayed at a smaller scale. *Displacement* avoids graphic interference by shifting apart features that otherwise would overlap or coalesce. A substantial reduction in scale, say from 1:25,000 to 1:1,000,000, usually results in an incomprehensibly congested collection of map symbols that calls for eliminating some features and displacing others. *Smoothing*, which also diminishes detail and angularity, might displace some points and add others to the list. A prime objective of smoothing is to avoid a series of abruptly joined straight-line segments. *Enhancement* adds detail to give map symbols a more realistic appearance. Lines representing streams, for instance, might be given typical meander loops, whereas shorelines might be made to look more coast-like. Enhanced map symbols are more readily interpreted as well as more aesthetic.

Point features and map labels require a slightly different set of generalization operators. Figure 3.2 illustrates that with point features, as with linear features, selection and displacement avoid graphic interference when too many close symbols might overlap or coalesce. When displacement moves a label ambiguously far from the feature it names, *graphic association* with a *leader line* or a numeric code might be needed to link the label with its symbol.

Figure 3.2. Elementary geometric operations in the generalization of point features and map labels.

Abbreviation is another strategy for generalizing labels on congested small-scale maps. *Aggregation* is useful where many equivalent features might overwhelm the map if accorded separate symbols. In assigning a single symbol to several point features, as when one dot represents twenty reported tornadoes, aggregation usually requires the symbol to either portray the "center of mass" of the individual symbols it replaces or reflect the largest of several discrete clusters.

Where scale reduction is severe, as from 1:100,000 to 1:20,000,000, *area conversion* is useful for shifting the map viewer's attention from individual occurrences of equivalent features to zones of relative concentration. For example, instead of showing individual tornadoes, the map might define a belt in which tornadoes are comparatively common. In highlighting zones of concentration or higher density, area conversion replaces all point symbols with one or more area symbols. Several density levels,

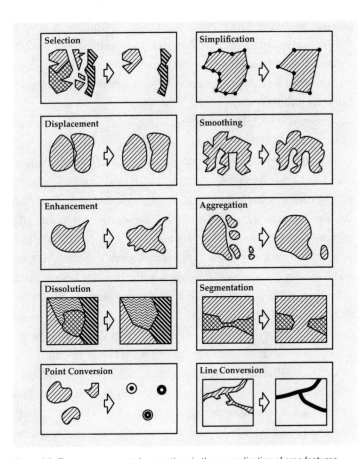

Figure 3.3. Elementary geometric operations in the generalization of area features.

perhaps labeled "severe," "moderate," and "rare," might provide a richer, less generalized geographic pattern.

Area features, as figure 3.3 demonstrates, require the largest set of generalization operators because area boundaries are subject to aggregation, point conversion, and all five elements of line generalization as well as to several operators unique to areas. Selection is particularly important when area features must share the map with numerous linear and point features. A standardized minimum mapping size can direct the selection of area features and promote consistency among the numerous sheets of a map series. For example, 1:24,000-scale topographic maps exclude woodlands smaller than one acre unless they are important as landmarks or shelter-belts. Soil scientists have used a less precise but equally pragmatic size threshold—the head of a pencil—to eliminate tiny, insignificant areas on soils maps.

Aggregation might override selection when a patch otherwise too small to include is either combined with one or more small, similar areas nearby or merged into a larger neighbor. On soils maps and land-use maps, which assign all land to some category, aggregation of two close but separated area features might require the *dissolution* or *segmentation* of the intervening area. A land-use map might, for example, show transportation land only for railroad yards, highway interchanges, and service areas where the right-of-way satisfies a minimum-width threshold. Simplification, displacement, smoothing, and enhancement are needed not only to refine the level of detail and to avoid graphic interference between area boundaries and other line symbols, but also to reconstruct boundaries disrupted by aggregation and segmentation.

Generalization often accommodates a substantial reduction in scale by converting area features to linear or point features. Line conversion is common on small-scale reference maps that represent all but the widest rivers with a single, readily recognized line symbol of uniform width. Highway maps also help the map user by focusing not on width of right-of-way but on connectivity and orientation. In treating more compact area features as point locations, point conversion highlights large, sprawling cities such as London and Los Angeles on small-scale atlas maps and focuses the traveler's attention on highway interchanges on intermediate-scale road maps. Linear and point conversion are often necessary because an area symbol at scale would be too tiny or too thin for reliable and efficient visual identification.

Comparing two or more maps showing the same area at substantially different scales is a good way to appreciate the need for geometric generalization. Consider, for instance, the two maps in figure 3.4. The rectangles represent the same area extracted from

Figure 3.4. Area near Northumberland, Pennsylvania, as portrayed on topographic maps at 1:24,000 (left) and at 1:250,000 enlarged to roughly 1:24,000 for comparison (right).

maps published at scales of 1:24,000 and 1:250,000; enlargement of the small-scale excerpt to roughly the same size as its more detailed counterpart reveals the need for considerable generalization at 1:250,000. The substantially fewer features shown at 1:250,000 demonstrate how feature selection helps the mapmaker avoid clutter. Note that the smaller-scale map omits most of the streets, all labels and all individual buildings in this area, and the island in the middle of the river. The railroad and the highway that cross the river are smoother and farther apart, allowing space for the bridge symbols added at 1:250,000. Because the 1:24,000-scale map in a sense portrays the same area in a space over a hundred times larger, it can show many more features in much greater detail.

How precisely are symbols positioned on maps? The US

Office of Management and Budget addressed this concern with the National Map Accuracy Standards, honored by the United States Geological Survey and other federal mapping agencies. To receive the endorsement that "This map complies with the National Map Accuracy Standards," a map at a scale of 1:20,000 or smaller must be checked for symbols that deviate from their correct positions by more than 1/50 inch. This tolerance reflects the limitations of surveying and mapping equipment and human hand-eye coordination. Yet only 90 percent of the points tested must meet the tolerance, and the 10 percent that don't can deviate substantially from their correct positions. Whether a failing point deviates from its true position by 2/50 inch or 20/50 inch doesn't matter—if 90 percent of the points checked meet the tolerance, the map sheet passes.

The National Map Accuracy Standards tolerate geometric generalization. Checkers test only "well-defined points" that are readily identified on the ground or on aerial photographs, are easily plotted on a map, and can be conveniently checked for horizontal accuracy; these include survey markers, roads and railway intersections, corners of large buildings, and centers of small buildings. Guidelines encourage checkers to ignore features that might have been displaced to avoid overlap or to provide a minimum clearance between symbols exaggerated in size to ensure visibility. In areas where features are clustered, maps tend to be less accurate than in more open areas. Thus Pennsylvania villages with comparatively narrow streets and no front yards would yield less accurate maps than, say, Colorado villages with wide streets, spacious front yards, and big lots. But as long as 90 percent of a sample of well-defined points not needing displacement meet the tolerance, the map sheet passes.

Maps that meet the standards show only *planimetric* distance— that is, distance measured in a plane. As figure 3.5 shows, a planimetric map compresses the three-dimensional land surface onto a two-dimensional sheet by projecting each point perpendicularly onto a horizontal plane. For two points at different elevations, the

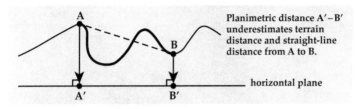

Figure 3.5. Planimetric map generalizes distance by the perpendicular projection of all positions onto a horizontal plane.

map distance between their "planimetrically accurate" positions underestimates both overland distance across the land surface and straight-line distance in three dimensions. Yet this portrayal of planimetric distance is a geometric generalization essential for large-scale flat maps.

The user should be wary, though, of the caveat "approximately positioned" or the warning "This map may not meet the National Map Accuracy Standards." In most cases such maps have been compiled from unrectified aerial photographs, on which horizontal error tends to be particularly great for rugged, hilly areas. Figure 3.6 shows the difference between the air photo's perspective view of the terrain and the planimetric map's representation of distances in a horizontal plane. Because lines of sight converge through the camera's lens, the air photo displaces most points on the land surface from their planimetric positions. Note that displacement is radially outward from the center of the photo, is greater for points well above the horizontal plane than for lower points, and tends to be greater near the edges than near the center. Cartographers call this effect "radial displacement due to relief," or simply *relief displacement*. An exception is the *orthophoto*, an air-photo image electronically stretched to remove relief displacement. An *orthophotomap*, produced from orthophotos, is a planimetrically accurate photo-image map. Chapter 12 examines a wider range of image maps.

For some maps, though, geometric accuracy is less important than linkages, adjacency, and relative position. Among the more

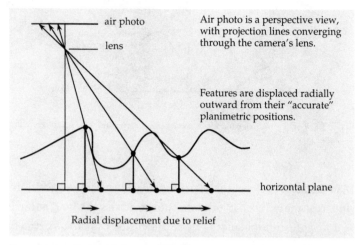

air photo

lens

Air photo is a perspective view, with projection lines converging through the camera's lens.

Features are displaced radially outward from their "accurate" planimetric positions.

horizontal plane

Radial displacement due to relief

Figure 3.6. A vertical aerial photograph (and any map with symbols traced directly from an air photo) is a perspective view with points displaced radially from their planimetric positions.

effective highly generalized maps are the linear cartograms por-traying subway and rapid transit systems. As in plate 1, showing the Washington, DC, Metro system, scale is relatively large for the inner city, where the routes converge and connect; stops in the cen-tral business district might be only four or five blocks apart, and a larger scale is needed here to accommodate more route lines and station names. By contrast, toward the fringes of the city, where stations are perhaps a mile or more apart, scale can be smaller because mapped features are less dense. Contrasting colors usually differentiate the various lines; the Metro system, in fact, calls its routes the Blue Line, the Red Line, and so forth, to enhance the effectiveness of its map. By sacrificing geometric accuracy, these schematic maps are particularly efficient in addressing the subway rider's basic questions: Where am I on the system? Where is my destination? Do I need to change trains? If so, where and to what line? In which direction do I need to go? What is the name of the station at the end of the line? How many stops do I ride before I get off? Function dictates form, and a map more "accurate" in the usual sense would not work as well.

Content

As geometric generalization seeks graphic clarity by avoiding overlapping symbols, content generalization promotes clarity of purpose or meaning by filtering out details irrelevant to the map's function or theme. Content generalization has only two essential elements, selection and classification. Selection, which serves geometric generalization by suppressing some information, promotes content generalization by choosing only relevant features. Classification, in contrast, makes the map helpfully informative as well as usable by recognizing similarities among the features chosen so that a single type of symbol can represent a group of similar features. Although all map features are in some sense unique, usually each feature cannot have a unique symbol. Even though some maps approach uniqueness by naming individual streets or numbering lots, these maps also use very few types of line symbols, to emphasize similarities among roads and property boundaries as groups. Indeed, the graphic vocabulary of most maps is limited to a small set of standardized, contrasting symbols.

Occasionally the "template effect" of standardized symbols will misinform the map user by grouping functionally different features. Standard symbols, designed for ready, unambiguous recognition and proportioned for a particular scale, are common in cartography and promote efficiency in both map production and map use. Traditional cartographers used plastic drawing templates to trace in ink the outlines of highway shields and other symbols not easily rendered freehand. Drafters would cut area and point symbols from printed sheets and stick them onto the map and apply dashed, dotted, or parallel lines from rolls of specially printed flexible tape. Currently, graphics software allows the mapmaker to not only choose from a menu of standardized point, line, and area symbols provided with the software but also design and store new forms, readily duplicated and added where needed. Consistent symbols also benefit users of the United States Geological Survey's series of thousands of large-scale topographic map sheets, all sharing a single graphic vocabulary. On highway maps, the key (or "legend") usually presents the complete set of symbols so that, at least while examining the map, the reader encounters no surprises. Difficulties arise, though, when a standard symbol must represent functionally dissimilar elements. Although a small, typeset annotation next to the feature sometimes flags an important exception—for instance, a section of highway that is "under construction"—mapmakers frequently omit useful warnings.

Generalized highway interchanges are a prime example of how information obscured by the template effect can mislead or inconvenience a trusting map user. The left panel of figure 3.7 is a detailed view of the interchange near Rochester, New York, between highways 104 and 590, as portrayed at 1:9,600 on a state-transportation-department map. Note that a motorist traveling from the east (that is, from the right) on NY 104 cannot easily turn north (toward the top of the map) onto NY 590. The upper-right portion of the left-hand map shows that the necessary connecting lanes from NY 104 were started but not completed. In contrast, the right panel shows how various commercial map publishers portray this interchange on their small-scale statewide highway maps. Two diamond-shaped interchange symbols suggest separate and equivalent connections with the eastward and westward portions of NY 104. Yet the large-scale map clearly indicates that a driver expecting an easy connection from NY 104 westbound onto NY 590 northbound must travel to the next exit west or south and then double back. Until the road builders complete their planned connecting lanes, such discrepancies between reality and art will frustrate motorists who assume all little diamonds represent full interchanges.

Effective classification and selection often depend on a mixture of informed intuition and a good working definition. This is particularly true for geologic maps and soils maps, commonly prepared by several field scientists working in widely separated places. A detailed description is necessary if two people mapping areas 100 miles apart must identify and draw boundaries for different parts of the same feature. These descriptions should also address the mapping category's internal homogeneity and the sharpness of its

Figure 3.7. Highway interchange near Rochester, New York, as portrayed on a detailed transportation-planning map (left) and on several commercial road maps (right).

"contacts" with neighboring units. In soils mapping, for instance, small patches of soil B might lie within an area labeled as soil A. This practice is accepted because these "inclusions" of soil B are too small to be shown separately and because the soil scientist cannot be aware of all such inclusions. Soils mapping, after all, is slow, tedious work that requires taking samples below the surface with a drill or auger and occasionally digging a pit to examine the soil's vertical profile. Map accuracy thus depends on the field scientist's understanding of the effects of terrain and geology (if known) on soil development, expertise in selecting sample points, and intuition in plotting boundaries.

That crisp, definitive lines on soils maps mark inherently fuzzy boundaries is unfortunate. More appalling, though, is the uncritical use in computerized geographic-information systems of soil boundaries plotted on "unrectified" aerial photos, which are subject to the relief-displacement error shown in figure 3.6. Like quoting a public figure out of context, extracting soils data from a photomap invites misinterpretation. When placed in a database with more precise information, these data readily acquire a false aura of accuracy.

Graphics software generally plays a positive role in map analysis and map display, the notion of "garbage in, garbage out" notwithstanding. Particularly useful is the ability of such software to generalize the geometry and content of maps so that one or two geographic databases can support a broad range of display scales. Large-scale maps presenting a detailed portrayal of a small area can exploit the richness of the data, whereas software-generalized smaller-scale displays can present a smaller selection of available features, suitably displaced to avoid graphic interference. Both the content and scale of the map can be tailored to the particular needs of individual users.

Software-generalized maps of land use and land cover illustrate how a single database can yield radically different cartographic pictures of a landscape. The three maps in figure 3.8 show a rectangular region of approximately 700 square miles (1,800 square kilometers) that includes the city of Harrisburg, Pennsylvania, above and slightly to the right of center. Software generalized these maps from a large, more detailed database that represents much smaller patches of land and describes land cover with a more refined set of categories. The generalization program used different sets of weights, or priorities, to produce the three patterns in figure 3.8. The map at the upper left differs from the other two maps because the computer was told to emphasize urban and built-up land. This map makes some small built-up areas more visible by reducing the size of area symbols representing other land covers. In contrast, the map at the upper right reflects a high visual preference for agricultural land. A more complex set of criteria guided generalization for the display at the lower left: forest land is dominant overall, but urban land dominates agricultural land. In addition, for this lower map the software dissolved water areas, which were discontinuous because of variations in the width of the river. These

Urban Land Dominant

Agricultural Land Dominant

Forest > Urban > Agriculture

Agricultural Land

Forest Land

Urban and Built-up Land

Water

20 miles

Figure 3.8. Land-use and land-cover maps generalized by computer from more de-tailed data according to three different sets of display priorities.

differences in emphasis might meet the respective needs and biases of demographers, agronomists, and foresters.

Generalized maps almost always reflect judgments about the relative importance of mappable features and details. The system-atic bias demonstrated by these generalized land-cover maps is not exclusive to software-generated maps; manual cartographers have similar goals and biases, however vaguely defined and unevenly applied. Through the consistent application of explicit specifica-tions, the generalization algorithm offers the possibility of a better map. Yet whether the map's title or description reveals these biases is an important clue to the integrity of the mapmaker or publish-er. Automated mapping allows experimentation with different sets of priorities. Hence software generalization should make the cartographer more aware of the multitude of choices, values, and biases. But just because a useful and appropriate tool is available does not mean the mapmaker will use it. Indeed, laziness and lack of curiosity all too often are the most important sources of bias.

The choropleth map (introduced as the right-hand elements of figures 2.13 and 2.14) is perhaps the prime example of this bias by default. Choropleth maps portray geographic patterns for regions composed of areal units, such as states, counties, and voting precincts. Usually two to six graytone symbols, on a scale from light to dark, represent two to six nonoverlapping categories for an intensity index such as population density or the percentage of the adult population voting in the last election. The breaks between these categories can markedly affect the mapped pattern, and the cautious map author tests the effects of different sets of class breaks. Mapping software can unwittingly encourage laziness by presenting a map based on a "default" classification scheme that might, for instance, divide the range of data values into five equal intervals. As a marketing strategy, the software developer uses such default specifications to make the product more attractive by helping the first-time or prospective user experience success. Too commonly, though, the naive or noncritical user accepts this arbitrary display as the standard solution, not merely as a starting point, and ignores the invitation of the program's pull-down menus to explore other approaches to data classification.

Different sets of categories can lead to radically different interpretations. The two maps in figure 3.9, for example, offer very different impressions of the spatial pattern of homes still lacking telephones in the northeastern United States in 1960, when minimal economic security demanded a landline. Both maps have three classes, portrayed with a graded sequence of graytone area symbols that imply "low," "medium," and "high" rates of phonelessness. Both sets of categories use round-number breaks, which mapmakers for some mysterious reason tend to favor. The map at the left shows a single state, Virginia, in its high, most deficient class, and a single state, Connecticut, in its low, most well-connected class. The casual viewer might attribute these extremes to Virginia's higher proportion of low-income African Americans and to Connecti-

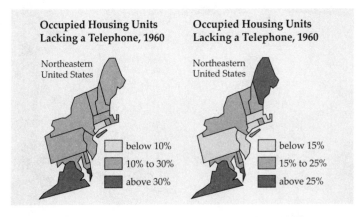

Figure 3.9. Different sets of class breaks applied to the same data yield different-looking choropleth maps.

cut's affluent suburbs and regard the remaining states as homogeneously "average." In contrast, the map at the right portrays a more balanced distribution of states among the three groups and suggests a different interpretation. Both states in the high category have substantial dispersed rural populations, and all four in the low category are highly urban and industrialized. Moreover, a smaller middle group suggests less overall homogeneity.

Machiavellian bias can easily manipulate the message of a choropleth map. Figure 3.10, for example, presents two cartographic treatments with substantially different political interpretations. The map on the left uses rounded breaks at 10 percent and 15 percent, forcing most states into its high, poorly connected category and suggesting a Northeast with generally poor communications. Perhaps the government is ineffective in regulating a gouging telecommunications industry or in eradicating poverty. Its counterpart on the right uses rounded breaks at 20 percent and 30 percent to paint a rosier picture, with only one state in the high group and eight in the low, well-served category. Perhaps government regulation is effective, industry benign, and poverty rare.

The four maps in figures 3.9 and 3.10 hold two lessons for the skeptical map reader. First, a single choropleth map presents only

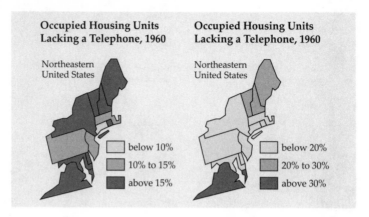

Figure 3.10. Class breaks can be manipulated to yield choropleth maps supporting politically divergent interpretations.

one of many possible views of a geographic variable. And second, the white lies of map generalization might also mask the real lies of the political propagandist.

Intuition and Ethics in Map Generalization

Small-scale generalized maps often are authored views of a landscape or a set of spatial data. Like the author of any scholarly work or artistic creation based on reality, the conscientious map author not only examines a variety of sources but also relies on extensive experience with the information or region portrayed. Intuition and induction guide the choice of features, graphic hierarchy, and abstraction of detail. The map is as it is because the map author "knows" how it should look. This knowledge, of course, might be faulty, or the resulting graphic interpretation might differ significantly from that of another competent observer. As is often the case, two views might both be valid.

Blunders That Mislead

Some maps fail because of the mapmaker's ignorance or oversight. The range of blunders affecting maps includes graphic scales that invite users to estimate distances on world maps, misspelled place-names, and graytone or color symbols changed by poor printing or poor planning. Maps can even be based on incompatible sources. By definition a blunder is not a lie, but the informed map user must be aware of cartographic fallibility and even a bit of mischief.

Cartographic Carelessness

Mapmakers are human, and they make mistakes. Although poor training and sloppy design account for some errors, most cartographic blunders reflect a combination of inattention and inadequate editing. If the mapmaker is rushed, if an employer assumes graphics software has made skilled cartographers unnecessary, or if no one checks and rechecks the work, missing or misplaced features and misspelled labels are inevitable.

Large-scale base maps, with background detail to which other information is added, have surprisingly few errors. A costly but efficient bureaucratic structure at government mapping agencies has usually guaranteed a highly accurate product, and commercial firms distributing base maps typically get much of the basic data from the government. Several layers of fact-checking and editing support technicians or contractors selected for skill and concern with quality. Making topographic maps is a somewhat tedious, multistep manufacturing process, and using outside contractors for compilation requires a strong commitment to quality control, buttressed by the bureaucrat's inherent fear of embarrassment. Blunders occasionally slip through, but these are rare.

Errors are more common on derivative maps—that is, maps compiled from other maps—than on basic maps compiled from air photos and other primary data. Artists lacking cartographic training and an appreciation of geographic details draw most tourist maps, and overworked media specialists produce most news maps. Omission and garbling are particularly likely when information is transferred manually from one map to another. Getting all the appropriate information from the large-scale base map onto a small-scale derivative map is not an easy chore. Several base maps might be needed, the compiler might not have a clear idea of what is necessary, or several compilers might work on the same map. Using another derivative map as a source can save time, but only at the risk of incorporating someone else's errors.

Map blunders make amusing anecdotes, and the press helps keep cartographers conscientious by reporting the more outrageous ones. In the early 1960s, for instance, the American Automobile Association "lost Seattle," as the Associated Press reported the accidental omission of the country's twenty-third largest city from the AAA's United States road map. Embarrassed, the AAA confessed that "it just fell through the editing crack" and ordered an expensive recall and reprinting.

Equally mortified was the Canadian government tourist office that omitted Ottawa from an airline map in a brochure intended to attract British tourists. The official explanation that Ottawa had not been a major international point of entry and that the map was compiled before initiation of direct New York City–Ottawa air service didn't diminish the irritation of Ottawa residents. The map included Calgary, Regina, and Winnipeg, and as an executive of the city's Capital Visitors and Convention Bureau noted, "Ottawa should be shown in any case, even if the only point of entry was by two-man kayak."

Faulty map reading almost led to an international incident in

1988, when the Manila press reported the Malaysian annexation of the Turtle Islands. News maps showing the Malaysian encroachment supported three days of media hysteria and saber rattling. These maps were later traced to the erroneous reading of an American navigation chart by a Philippine naval officer who mistook a line representing the recommended deepwater route for ships passing the Turtle Islands for the boundary of Malaysia's newly declared exclusive economic zone.

Although map blunders provoking outrage between minor powers make amusing anecdotes, inaccurate maps in a war zone can be deadly. The American Civil War illustrates the effect, on both sides, of wildly conflicting topographic maps and inadequate numbers of trained topographic engineers and geographers. In 1862, for instance, the Union army planned a swift defeat of the Confederates by capturing their capital, Richmond. But unexpected obstacles slowed the northern army's advance after General McClellan's staff based battle plans on inaccurate maps. A lack of good maps also plagued Confederate forces, who were unaware of strategic advantages that would have allowed them to overwhelm McClellan's retreating army.

Modern warfare is especially susceptible to bad maps. In May 1999, during a United Nations (UN) peacekeeping mission in the former Yugoslavia, an American smart bomb destroyed the Chinese embassy in Belgrade because of faulty targeting information—a failure to match the right street address with the right building. Tragically, two months earlier, twenty people had died in northern Italy when a Marine Corps jet struck a ski lift that was missing from a map of vertical obstructions. In January 2005, a fast-moving US nuclear submarine hit an uncharted undersea mountain 350 miles south of Guam. Submarines are particularly vulnerable because the need to hide from the enemy precludes the use of active sonar, which might have warned of the obstruction.

Then there was the 1983 invasion of Grenada by United States troops and their Caribbean allies. The only cartographic intelligence distributed to troops sent to rescue American medical stu-

Figure 4.1. Reconstruction of a journalistic misrepresentation (right) of the size and location of Grenada's smaller islands, as presented in a book published shortly after the 1983 invasion. The left and upper portion of the figure portrays Carriacou and Petite Martinique correctly.

dents consisted of hastily printed copies of a few obsolete British maps and a tourist map with a military grid added. An air attack destroyed a mental hospital not marked on the maps. Another air strike, ordered by a field commander using one set of grid coordinates but carried out by planes using a map with another grid, wounded eighteen soldiers, one fatally.

Journalists' and social scientists' accounts of the invasion added further cartographic insult. In addition to using misplaced symbols and misspelled place-names, one group of coauthors (or their freelance illustrator) distorted the sizes and relative positions of the Grenadian state's two smaller members, the islands of Carriacou and Petite Martinique. As the lower-right part of figure 4.1 illustrates, their regional locator map made these islands much smaller than their true relative sizes and moved them much closer to the main island. These errors probably originated in careless compilation from an official map on which an inset showed the

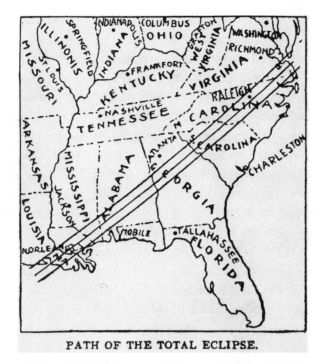

PATH OF THE TOTAL ECLIPSE.

Figure 4.2. On May 28, 1900, the *Cortland* (NY) *Evening Standard* printed this hand-drawn map, received from a feature syndicate, with the name Illinois misspelled.

smaller islands at a smaller scale than the main island. Since an earlier derived map had the same errors, the authors apparently based their map on a faulty source, which lax editorial checking obviously failed to detect.

The news media have their own cartographic glitches. Errors on news maps reflect both the minimal cartographic knowledge of their creators and the high-pressure, deadline-driven environment in which news is gathered, processed, and published or posted. This is not a new development. Many news maps distributed around 1900 had the crude boundaries and careless spelling exemplified by figure 4.2, a map accompanying a syndicated story about a coming eclipse; note the extra *n* in the label for Illinois. Hastily drawn news maps have also annexed Michigan's northern peninsula into Wisconsin, Virginia's eastern counties on the Delmarva Peninsula into Maryland, and both North Korea and South Korea (and part of China) into the former Soviet Union.

Illustration software, a more recent impetus for journalistic cartography, makes it easier for reporters and editors to alter decent-looking base maps and inadvertently eliminate features and misplace symbols and labels. Typical examples include adding Finnish territory to the Soviet Union on the map decorating a *New York Times* article on Canadian–Soviet relations and switching the labels identifying New Hampshire and Vermont on a Knight-Ridder Graphics Network map of areas in the United States affected by drought.

Perhaps more annoying are blunders on road maps and street maps—electronic or paper—particularly when the place you are trying to find is missing, misplaced, misindexed, mislabeled, or badly misshaped. That these errors are not more common is surprising, though. Publishers of street and highway maps must manage a complex, constantly changing database and produce a low-cost, enormously detailed product for largely unappreciative consumers in a highly competitive market. Perhaps because oil companies distributed free road maps for several decades until the early 1970s, and because state and local tourist councils perpetuate the free travel map, the American map buyer has little appreciation of the well-designed, highly accurate maps that the European map user is conditioned to respect, demand, and pay for.

Blunders on street maps reflect how the maps are made. Basic data for established parts of an area can be found on large-scale topographic maps published by the Geological Survey. These maps are in the public domain and can be copied freely, but their publication scale of 1:24,000 allows room for few street names, and many map sheets are ten years or more out of date. Thus, mapmakers have turned to maps maintained by city and county engineering and highway departments for street names as well as for new streets and other changes. Personnel who are responsible for copying street alignments, typesetting, and type placement are sometimes inexperienced or inattentive, and editing is not always thorough. A file of customers' complaints can be a useful though

Figure 4.3. Two paper streets in Syracuse, New York, as shown on a municipal street map (left panel; see Garden Street and Pinnacle Street just to the left of the center of the map) and the corresponding area as represented on a New York State highway planning map. (In the left panel a tiny arrow points to block-long Pinnacle Street, which is parallel to Smith Street.)

not fully reliable source of corrections for the next edition. Some publishers have sent drafts of their maps to planning and engineering departments for comment and have unrealistically depended on overworked civil servants for additional editing. Similar problems affect digital compilation.

Unfortunately, many official city maps show the rights-of-way of approved streets that were never cleared, graded, or paved. Often a planned but otherwise fictitious street persists on the city engineer's map until it is officially deleted, as when a developer petitions the planning board to build across the right-of-way or a homeowner attempts to buy the adjoining strip of land. Street-map publishers who compile only from municipal maps are likely to pick up a number of "paper streets" like Garden Street and Pinnacle Street in figure 4.3, found on an official map for Syracuse, New York. Geological Survey and New York State quadrangle maps of the area don't show these streets, but the mapmaker might conveniently assume that the "official" map provided by local government is more likely to be accurate than a map from Washington or the state capital.

Deliberate Blunders

Although none dared to talk about it, publishers of street maps often turned to each other for street names and changes. The euphemism for this type of compilation was "editing the competition," but the legal term is copyright infringement—if you cribbed from a single source and got caught. To be able to demonstrate copyright infringement in court, and possibly to enjoy a cash settlement by catching a careless competitor in the act, some map publishers deliberately falsified their maps by adding "trap streets." As deterrents to the theft of copyright-protected information, trap streets were usually placed subtly, in out-of-the-way locations unlikely to confuse or antagonize map users. Map publishers were understandably reluctant to talk about this deliberate fudging, which died off after 1997, when a federal district court ruled that false facts like trap streets were facts, nonetheless, and thus not eligible for copyright protection.

Map drafters having fun are another source of cartographic fiction. Michigan's state highway map for 1979, for instance, included two fictitious towns reflecting the traditional football rivalry between the University of Michigan and Ohio State University. As figure 4.4 shows, the cartographic culprit clearly was not only a Michigan fan but a loyal citizen of Michigan, perpetrating place-name pollution only in neighboring portions of Ohio—and perhaps thinking that the editor would be less careful in checking out-of-state features. Toledo's new eastern suburb "goblu" is a slightly compacted version of the familiar cheer of fans rooting for the Michigan Blue (a nickname based on the primary school color), and the new town "beatosu," north of Burlington and south

Figure 4.4. Fictitious towns "goblu" (above, to the right, below Bono) and "beatosu" (toward the bottom, above Burlington) on the 1979 Michigan highway map reflect an unknown mapmaker's support for the University of Michigan football team (the Blue) over its traditional rival, Ohio State University (OSU).

of the Ohio Turnpike, reflects the Blue's principal annual gridiron goal, defeating OSU.

A more personal example of creative cartography is Mount Richard, which, in the early 1970s, suddenly appeared on the continental divide on a county map prepared in Boulder, Colorado. Believed to be the work of Richard Ciacci, a draftsman in the public-works department, Mount Richard was not discovered for two years. Such pranks raise questions about the extent of yet-undetected mischief by mapmakers reaching for geographic immortality.

Distorted Graytones: Not Getting What the Map Author Saw

Printing can radically alter the appearance of a map, and failure to plan for the distorting effects of reproduction can yield thousands of printed maps that look quite different from the original artwork. Map images can lose clarity when artwork is photographed, scanned, or transferred (photographically or electronically) to a printing plate, or when ink is transferred from the printing plate onto the paper. An underinking printing press or the overexposure or underdevelopment of the photograph yields a faint image, from which fine type or fine area patterns might have disappeared. An overinking press or the underexposure or overdevelopment of the photograph might fill in the corners and tightly closed loops of small type and can noticeably darken some graytone area symbols. Usually the culprit is an overinking printing press, which fattens image elements through a malfunction called *ink spread*.

Fine dot screens used as area symbols are especially vulnerable to ink spread, and the map designer who underestimates the effect of ink spread on screens with dots spaced 120 or more to the inch (47 or more per centimeter) risks medium grays that turn out black and choropleth maps on which low becomes high and high becomes low. Figure 4.5 uses greatly enlarged views of two hypothetical graytone area symbols to illustrate this effect on a relatively fine 150-line screen (59 lines per centimeter), with dots spaced 0.007 inch (0.169 mm) apart. In this simulation a small amount of ink spread that increases the radius of each dot by 0.001 inch (0.025 mm) raises the black area of a 20 percent screen to 49 percent and that of an 80 percent screen to 96 percent. Screens for original graytones of less than 50 percent black grow darker because ink added around the edges of tiny black dots makes the dots larger, in some

Figure 4.5. Enlarged diagram showing the effects of 0.001 inch of ink spread on 20-percent (above) and 80-percent (below) 150-line graytone dot screens.

cases causing them to coalesce. Screens for original graytones of more than 50 percent black grow darker because ink added on the inside edges of tiny clear dots on a black background makes the clear dots smaller, in some cases completely filling them in. The cautious mapmaker adjusts screen texture to printing quality. The same amount of ink spread applied to moderately coarse, 65-line screens (26 lines per centimeter) would increase the 20 percent screen to only 32 percent black and the 80 percent screen to only 89 percent black.

Electronic graphics introduce further complications when a map author who is pleased with the image on a computer screen ignores the distortion that might occur when the map is printed with ink or toner, reproduced as a poster with an ink-jet printer, projected onto a screen in an auditorium, or viewed on a tablet computer. Colors might not look the same on a different display.

Another distortion is the so-called Xerox effect, which arises when a black-and-white photocopier destroys readily perceived differences between solid colors. Light blue might no longer be distinguishable from yellow, for example, and dark blue and bright red can appear similarly dark. Moreover, light or medium blue lines might wash out or break up. Similar distortions can occur when a color map distributed online is reproduced on a black-and-white laser printer, and maps with otherwise reliable sets of gray-tones are easily undermined by weak toner cartridges.

Temporal Inconsistency: What a Difference a Day (or Year or Decade) Makes

Maps are like milk: their information is perishable, and it is wise to check the date. But even when the map author provides one, the date might reflect the time of publication, not the time for which the information was gathered. And when the map was compiled from more than one source or through a long, tedious field survey, the information itself might be so temporally variable as to require not a single date but a range of dates. Particularly troublesome is the carefully dated or current-situation map that shows obsolete features or omits more recent ones. These errors might be few and not readily apparent; a map that is 99.9 percent accurate easily deceives most users.

Inaccurately dated or temporally inconsistent maps can be a particular hazard when the information portrayed is volatile. A map of past geological conditions might inaccurately estimate its dates by thousands or even millions of years and still be useful, whereas the temperature and pressure observations used to prepare weather maps must be synchronized to within an hour or less. Moreover, maps forecasting weather patterns must state accurately the date and time for the forecast.

Historians in particular should be skeptical of dates on maps. Medieval maps, for instance, can cover a much broader range of time than a single year-date suggests. As historian of cartography

David Woodward noted, medieval *mappaemundi* (world maps), instead of providing an accurate or perceived image of the earth at an instant of time, often "consist of historical aggregations or cumulative inventories of events that occur in space." For instance, the famous Hereford map, named for the British cathedral that owns it, was compiled around 1290 from a variety of sources. Its place-names present an asynchronous geography ranging from the fourth-century Roman Empire to contemporary thirteenth-century England.

Conscientious users of modern maps read whatever fine print the mapmaker provides. A large-scale topographic map released in year N with a publication date of year N – 2, for instance, might be based on air photos taken in year N – 3 or N – 4 and might have been field checked in year N – 2 or N – 3. But field checking might not detect all significant changes, and maps of areas undergoing rapid urban development oftentimes are appallingly obsolete. Moreover, derived maps without fine print can be the cartographic equivalent of snake oil. Because of publication delays and slow revision, "new" derived maps may well be ten years or more out of date. Or they may be only four years out of date in some areas, and ten years or more in others.

A temporally accurate map need not focus on a single date or include only features that currently exist or once existed. Planners, for instance, need maps to record their previous decisions about future projects so that new decisions do not conflict with old ones. Fairfax County, Virginia, learned that inaccurate planning maps can have costly consequences. Because of inadequate mapping, a developer was authorized to start building a subdivision in the path of a planned limited-access highway. Buying the seventeen affected lots cost county taxpayers $1.5 million. The same highway also required the county to buy and raze five new homes in another subdivision. After these two embarrassing incidents, county officials set up a special mapping office.

Temporal consistency is also troublesome for users of statistical maps produced from census and survey data. Choropleth maps and other data maps sometimes portray a ratio or other index that compares data collected for different instants or periods of time. Usually the temporal incompatibility is minor, as when the Census Bureau computes per capita income by relating the population count for April 1 of a round-number year to the area's total income for the previous year—getting an accurate count of the population and honest and reliable estimates of personal income is far more problematic. One must beware, though, of indexes that relate mid-decade survey data and beginning-of-decade census tabulations, for example, by dividing 2015 income totals by 2010 populations. Migration can significantly distort ratios based on asynchronous data, particularly for small areal units such as suburban towns, which can grow by 500 percent or more in ten years.

International data based on inconsistent definitions as well as asynchronous censuses or surveys can yield highly questionable maps. Worldwide maps of poverty, occupational categories, and the proportion of the population living in urban areas are inherently imprecise because of significant international differences in the relevant definitions. World maps based on statistical data are particularly suspect—whatever validity they have arises not only from knowing adjustment by scholars or United Nations officials but also from broad, very general categories that tend to mask spurious differences within the groups of more and less developed nations.

Incompatible Data: Mismatched Ingredients Can Spoil a Stew

Maps based on electronic data files can be highly erroneous when several sources contributed the data and the user or compiler lacked the time or interest to verify their accuracy. Obviously inaccurate data from careless, profit-driven firms have unpleasantly surprised purchasers of street-network information, and even data from reliable vendors occasionally have infuriating errors, as when roads and boundaries captured from adjoining map sheets fail to align at the common edge. Without careful editing, streets and

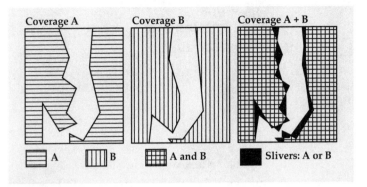

Figure 4.6. Spurious sliver polygons result from the overlay of two inaccurately digitized coverages A and B.

streams are more easily omitted or misplaced in a computer database than on a paper map. Electronic data are incomplete without metadata (data about data) that address quality assurance and help data users determine the compatibility of diverse features that might be combined.

Transfer from a paper map to an electronic format invites inconsistencies when multiple sets of features, or "coverages," are available for map overlay in a geographic information system. Meaningless results are likely when features extracted from different maps of the same area are misaligned, perhaps because their source maps have different projections. Overlays of spatially or temporally incompatible data—for example, the overlay of two closely related coverages that were collected separately—can yield slivers and other spurious polygons, as the example in figure 4.6 demonstrates. Moreover, the software used to process the information might also be flawed. Although software errors can be blatant, subtle programming errors could lead to disastrous decisions; this possibility should encourage the user to carefully explore unfamiliar data and software. "Garbage in, garbage out" is a useful warning, but sometimes you can't tell the data are garbage until they have been used for a while.

CHAPTER 5

Color: Attraction and Distraction

Color is a cartographic quagmire. Color symbols can make a map visually attractive as well as fulfill the need for contrast on road maps, geological maps, and other maps with many categories. Yet the complexity and seductiveness of color overwhelm many mapmakers, and countless maps in computer-graphics demonstrations, business presentations, and news media reveal a widespread ignorance of how color can help or hurt a map. Persons unaware of the appropriate use of color in cartography are easily impressed and might accept as useful a poor map that merely looks pretty.

Technological change accounts for much of the misuse of color on maps. Before the 1980s color printing was expensive and seldom thoughtlessly used, and color maps were comparatively rare. Advances since then in computer hardware and software and in online mapping have encouraged a fuller use—and abuse—of color. Color monitors and printers have made color effortlessly available to the amateur mapmaker; run-of-press color lithography encouraged increased use by the cartographically illiterate commercial artists responsible for most news illustration; and high-resolution screens on laptops, tablets, and mobile devices have made color maps pervasive if not ubiquitous online. Moreover, many viewers and readers expect maps with richly contrasting hues, even when black-and-white or more subdued symbols might be more readily and reliably decoded. This chapter briefly explains the nature of color and examines how graphic logic, visual perception, and cultural preferences affect the use of color on maps.

The Phenomenon of Color

As a biophysical phenomenon, color is a sensory response to electromagnetic radiation in a narrow part of the wavelength spectrum between roughly 0.4 μm and 0.7 μm, called the "visible band." (One micrometer [μm] is one-millionth of a meter.) The eye cannot see shorter-wave radiation, such as ultraviolet light (10^{-1} μm) or gamma rays (10^{-6} μm), nor can it sense longer-wave energy, such as microwave radiation (10^5 μm) and television signals (10^8 μm). Yet within the visible band the eye and brain readily distinguish between wavelengths associated with the hues we call violet, blue, green, yellow, orange, and red, as in the left half of figure 5.1. White light is a mixture of all these wavelengths, but a rainbow or prism can refract white light and sort its constituent colors into this familiar spectral sequence.

As a perceptual and graphic-arts phenomenon, color has three dimensions: hue, value, and saturation. *Hue*, related to the wavelength of electromagnetic radiation, is the most obvious gauge of color; when most people think of color, they think of hue (plate 2). *Value*, which refers to a color's lightness or darkness, applies to both hues and shades of gray. *Saturation*, also called *chroma*, refers to a color's intensity or brilliance; a medium blue, for instance, might range in saturation from a pure, 100 percent–saturated, strong blue down through moderate blue, weak blue, and bluish gray, to a 0 percent–saturated medium gray. Graytones, also called *achromatic color*, have zero saturation.

Color theorists often invoke the HVS color-space diagram shown in the right half of figure 5.1 to describe the three-dimensional relationship of hue, value, and saturation. A scale similar to the visible band but bent into a circle that joins dark red with dark purple represents the range of hues. This circle of hues, or *color wheel*, is centered on and perpendicular to a vertical axis of graytone values, which range from black at the bottom to white at the top. Saturation, the third dimension, measures a color's distance outward from this graytone axis toward one of the pure hues on the

Colors of the
Visible Spectrum

Three-Dimensional
HVS Color Space

Figure 5.1. The visible part of the electromagnetic spectrum (left) shows the relation between wavelength and hue. The HVS color space (right) links hue with the other two dimensions of color: value and saturation.

color circle; a fully saturated blue, for instance, would lie farther from the value axis than a weak blue.

As visual variables, hue and value play very different roles. Although we perceive blue, green, and red as somehow different, we readily organize various shades of red along a scale from light red to dark red. Designers cannot manipulate chroma as easily as hue and lightness, and they rarely use it as a visual variable, at least not consciously.

Colors on maps usually are mixtures of primary colors. Computer screens, mobile devices, and video projectors generate a wide variety of hues by mixing the *additive primary colors* of red, green, and blue. The left half of plate 3 demonstrates some of the mixed colors occurring when red, green, and blue spotlights produce overlapping circles on a white, reflective stage in an otherwise dark theater. Yellow occurs when green light is added to red light, cyan is a hue between blue and green, and magenta is a reddish purple reflecting the addition of red and blue. Overlapping red, green, and blue circles of light on an electronic display would yield the same effect. Note that white occurs at the center, where all three circles overlap. White light results from mixing red, green, and blue wavelengths of roughly equal intensity. Outside the area covered by one or more spotlights, the stage appears black, indicating the absence of color.

Recognition of white light as the summation of red, green, and blue makes it easy to understand the *subtractive primary colors*. Yellow, magenta, and cyan are called subtractive primaries because each represents the subtraction of one of the additive primary colors from white light. For example, a white card that is dyed yellow will absorb blue light from white light and reflect the remaining mixture of red and green light, which the eye perceives as yellow. That is, white (which is red + green + blue) minus blue (removed by the yellow dye) leaves red plus green, which is yellow. Similarly, a magenta dye removes green wavelengths from a white-light mixture, and a cyan dye removes red light from white light. Printed maps have color because dyes on the paper are selectively absorbing components of white light, whereas maps on electronic displays have color because each pixel has a trio of independently triggered phosphor "dots" that are able to emit electromagnetic radiation in the red, green, or blue parts of the spectrum.

As with the additive primary colors, mixing appropriate amounts of yellow, magenta, and cyan can generate other hues. The right half of plate 3 shows the pattern of colors produced by overlapping circles dyed with the three subtractive primaries. Note that an additive primary color occurs where only two circles overlap and that black (the absence of color) occurs where all three overlap. Red, for instance, occurs where yellow and magenta dyes overlap because subtracting blue and green respectively from white light leaves only red. Black occurs in the triple-overlap center zone because the three dyes together subtract all of the additive primaries.

Primary colors simplify both electronic display and printing. Colors on computer screens are based on the additive primaries and three separate arrays of thin, closely spaced phosphor dots or lines, 50 or more to the inch. (If this concept is unfamiliar, look very closely at a TV or monitor—but not for long!) In an area of the screen, each set of intermingled red, green, and blue dots or lines can be excited to emit light at a different intensity. If 64 different intensity levels are provided for each color, the RGB (Red-Green-Blue) color monitor can in theory produce 262,144 (64^3) different colors. A three-dimensional diagram called the RGB color cube, as in the left half of plate 4, represents the range of color possible with an RGB color monitor; any point within the cube represents a unique color, defined by unique coordinates for the cube's three axes. But some graphics systems restrict the user to a *palette* with a much smaller, more manageable number of discrete choices.

Sets of intermingled dots are also the basis for printed colors, but these "screened" dots vary in size rather than in intensity and absorb rather than emit light. A technique called *process printing* depends on a reflective, light background and uses transparent inks of the three subtractive primary hues to absorb varying amounts of red, green, and blue. Thus an area with comparatively large yellow dots and small cyan dots printed on white paper, as in the right half of plate 4, will absorb a moderate amount of blue light as well as some red light and will appear green. Because the dots are small and closely spaced—a typical screen arranges the dots in a fine grid with rows of 50 to 150 tiny dots *per inch*—our eyes ignore the screen's texture and orientation but mix its dyes. In process printing, a fourth impression is made using black ink to assure a strong black for type and fine lines.

Electronic color is not uniformly reliable. Although digital encoding has largely eliminated differences in television reception that once undermined the designer's intent, glare from a bright light or the sun can still be troublesome. For this reason, televised maps still tend to have highly contrasting hues, even when the data represent differences in intensity rather than kind. Although monitors from different manufacturers are more consistent than pre-digital TV sets, a video projector might not produce colors identical to those on a monitor connected to the same computer.

Printing can also thwart the map author's intent, especially where the printing press transfers extra ink to the paper, as with ink spread, discussed in chapter 4. Moreover, a map reproduced on a color inkjet printer can look quite different from its counterpart created with colored toners on a laser printer. Disparities are particularly acute when the ink or toner for one of the device's hues is exhausted or nearly so.

Color on Maps

How can you tell when a map in color might be misleading? The wary map user must first ask whether the map uses color—that is, hue—to portray differences in intensity or differences in kind. Soils maps, geologic maps, climatic maps, vegetation maps, zoning maps (see fig. 7.1 and plate 5), land-use maps, road maps, and many other types of maps showing a variety of features can benefit from contrasting hues, provided that somewhat similar hues represent somewhat similar features and radically different hues represent radically different features. Be suspicious, though, when contrasting hues attempt to show differences in intensity on choropleth maps (discussed in chap. 11).

Hue differences usually fail at portraying differences in percentages, rates, median values, and other intensity measures because spectral hues have no logical ordering in the mind's eye. Consider the experiment described in figure 5.2, a mass of cards in seven different hues, all with the same lightness and chroma. If ten people were asked separately and without collaboration to choose seven cards of different colors and arrange them in order from low to high, the odds are they would produce ten different sequences. Some might order them from green to red, some from blue to red, and some might attempt to reproduce a rainbow sequence. Two people might even agree on the same series but disagree on its direction from low to high. The experiment would demonstrate clearly that there is no simple, readily remembered and easily used sequence of hues that would obviate a map reader's need to refer back and forth repeatedly between map and key. Color symbols can be used on choropleth maps, but not conveniently. The use of spectral hues to portray intensity differences is a strong clue that

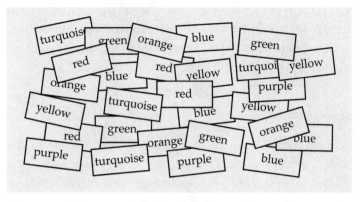

Figure 5.2. People tend to be unable to sort a variety of spectral hues into a single, consistent sequence of colors.

the mapmaker either knows little about map design or cares little about the map user.

Not all use of color for choropleth maps is confusing or undesirable. Value differences and hues coexist nicely in some single-sequence, part-spectral color scales, such as the sequence from light yellow to black in the upper map in plate 6. Note that this progression from light yellow to brown-yellow to brown to dark brown to black has a consistent, logical, and readily comprehended ordering from light to dark. Not only does this sequence of yellows, browns, and black show the pattern of high and low values as effectively as the graytones in figure 5.3, but the color map is more aesthetically appealing. In contrast, although young children and some adults might be attracted to the lower map in plate 6, most users will find its full spectral scale of primary hues confusing, complex, and comparatively difficult to decode, especially for intensity data, like median age or population density, for which darker-means-more is easy to understand and apply.

Plate 7 demonstrates the variety of color scales found on choropleth maps. The gray scale in the upper left might, for example, adopt a single hue, as at the lower left, without any loss in consistency, graphic logic, or ease of use. A partial spectral scale based

Consistent Gray Scale

0 - 9

10 - 19

20 - 29

30 - 39

40 - 50

Figure 5.3. A logical, consistent sequence of graytones describes intensity variations more reliably than a complex, graphically illogical sequence of spectral hues (see plate 6, bottom).

on yellow, orange, and red, as in the middle of the top row, can be as consistent, convenient, and visually appealing as the yellow-brown-black sequence in plate 6; yellow to green to dark blue provides another common single-sequence, part-spectral color scale useful on choropleth maps. Among the readers of color weather maps, the more graphically complex spectral scale at the upper right not only draws on useful associations of blue with cold and red with hot, but also on the reinforcement of daily exposure to the same color scheme; yet for most choropleth maps this full-spectral sequence is almost as awkward and troublesome a way to introduce color as the multi-hue, nonspectral sequence at the lower right. In comparison, the double-ended scale in the middle of the lower row uses contrasting hues to support a readily comprehended graphic logic, useful for maps showing both positive and negative rates of change. In this scheme, the colder blues represent significant losses or declines, the warmer reds indicate major gains or increases, neutral grays represent minor change, and weak blues and reds show moderate decreases and increases, respectively. When reinforced on a succession of maps, this color scale can be a powerful tool for mapping historical economic data.

Because colors distinctly different in hue can be highly similar in value, map users should be suspicious of maps received by fax or reproduced on a black-and-white photocopier. Dark red and dark blue can copy as black or dark gray, for instance, while yellow might turn white. Map authors too should be mindful of the "Xerox effect" and anticipate electronic distortion of maps that contain complex colors. In the spirit of government reports with "intentionally left blank" stamped on otherwise empty pages, maps might include a small note stating whether they were first printed in color or monochrome.

The conscientious map user must also be wary of the deliberate or inadvertent use of color to make a feature or proposal appear attractive or unattractive. People respond emotionally to some colors, such as blue and red, and some of these responses are common and predictable enough to be tools of the cartographic propagandist. And even if no deliberate manipulation is intended, because of embedded emotions or culturally conditioned attitudes some colors carry subtle, added meaning that could affect viewers' interpretation of a map or their feelings about the map or the elements it portrays.

Color preferences vary with culture, life cycle, and other demographic characteristics. For instance, men tend to prefer orange to yellow and blue to red, whereas women favor red over blue and yellow over orange. Preschool children like highly saturated colors, such as bright red, green, and blue, whereas affluent middle-aged adults generally prefer subtler, pastel shades. Among the spectral colors, North American adults seem to prefer blue and red to green and violet, and green and violet to orange and yellow. Least appreciated is a vomit-like greenish yellow.

Preferences also extend to groups or ranges of colors. A range of greens and blues, for instance, is generally preferable to a range of yellows and yellow-greens. Among people who like earth tones, a yellow-brown sequence would be attractive.

Little is known about the effects on map users of a variety of subjective reactions to color. Most colors, in fact, relate to several concepts, favorable and unfavorable. Because red, for instance, is associated with fire, warning, heat, blood, anger, courage, power, love, material force, the British Empire, Communism, and states that vote Republican, its effect probably depends very much on context. A right-wing political group might conveniently paint the former Soviet Union, Cuba, or China red on a world map, for instance, whereas a marketing manager might use a map with red target areas to focus the attention of corporate managers. Plate 8 illustrates how the addition of progressively redder, more intense tints makes a forceful propaganda map (fig. 8.16) even stronger.

Similarly, black might connote mourning, death, or heaviness, whereas blue can suggest coldness, depression, aristocracy, submissive faith, or states that vote Democrat. White might suggest cleanliness or sickness, and green can relate to envy, compassion, concern for the environment, or the Irish. Yellow's subjective message, if any, clearly depends on context: its possible use as a symbol of weakness and cowardice contrasts with its almost equally strong association with cheerfulness and power.

Some designers find color to be a clever or an obvious reinforcement to pictorial symbols. Given a golden tinge and used over a dollar sign, for instance, yellow reinforces an icon of wealth. Other examples of color-enhanced icons are green shamrocks, symmetrical red crosses on ambulances or hospitals, vertical black crosses for cemeteries or churches, and lemon-yellow cars with flat tires.

Maps portraying environmental hazards often borrow the familiar red-yellow-green sequence of traffic-light hues, as in the landslide-hazard map in plate 9. This sequence is highly effective, at least among map viewers who drive, because of continually reinforced associations of red with danger, yellow with a need for caution, and green with lower risk. Even so, reliable use of these colors requires the map author to explain the metaphor, perhaps with a stoplight icon in the map key.

Color's effectiveness as a map symbol might conflict with or reinforce its role as a landscape metaphor. For several centuries, cartographers have exploited and encouraged such associations as green with vegetation, blue with water, red with high temperatures, and yellow with a desert environment, and where the context is correct and appropriate, these associations promote efficient decoding. But some caution is warranted, for the blueness of the water might exist largely in the minds of wishful environmentalists, self-serving tourist operators, and gullible map readers.

A cartographic scheme called *hypsometric tints* can be particularly misleading to map users unaware of its focus on elevation. Plate 10 illustrates a common practice for portraying relief on general-purpose wall maps and atlas maps: a multi-hue color scale with five or more steps ranging from medium green for low elevations through yellow to orange or white for the highest elevations. Although widely used, hypsometric tints have not been fully standardized, and a variety of variations and modifications occur, including dark green for elevations below sea level and a dark, rusty brown for higher elevations. Confusion is likely if the reader associates white with snow, green with abundant vegetation, and yellow or brown with desert—much of the world's tundra is close to sea level, many lowland areas are deserts, and many upland areas have substantial forests or grasslands.

Conscientious users of color maps must also be wary of *simultaneous contrast*, the eye's tendency to perceive a higher degree of contrast for juxtaposed colors. When a light color is surrounded by a dark color, as in plate 11, simultaneous contrast will make the light color seem lighter and the dark color seem darker. Thus a medium gray or blue surrounded by darker symbols will appear lighter, whereas its sample in the map key is surrounded by white and appears darker. This effect can be particularly troublesome on geological maps and on some environmental maps with many categories, especially when only slightly different colored symbols represent markedly different categories.

Another perceptual effect that causes confusion between colors on the map and colors in the key is the tendency for large patches of color to look more saturated than small patches of the same color. For example, a large area of moderate green might appear to match a small sample of bright green in the key. On maps with many categories and varied colors, the good cartographer provides a redundant stimulus, such as alphanumeric codes or patterned area symbols overprinted in black, to help the conscientious map reader use the map key accurately.

Map viewers searching a color map for a place or feature must watch for inadvertent camouflage. Because geography can juxtapose odd colors, poor planning by the map author often leads to poor contrast between type and point symbols and their background, including such atrocities as yellow type on a white background and purple or blue type on a black background. Both combinations produce labels that are easily overlooked and difficult to read. Yet yellow works well against black, and purple can be highly legible on white, especially if lighting is poor. Type is more likely to be illegible when a label must cross both light and dark backgrounds.

People with impaired color vision—roughly 5 percent of the population, and mostly male—must be particularly cautious. Because the kind and degree of color blindness varies, there is no universal fix. Since difficulty in distinguishing red and green is the most common variety, textbooks advise map authors to use red-blue or blue-green pairings, rather than red-green, to distinguish a map's most different categories. Color blindness can also undermine the legibility of type and line symbols, particularly if there is poor contrast in lightness with a label's or a line's surroundings.

Lacking experience with electronic displays and additive colors, amateur mapmakers sometimes try to mimic printed maps on an electronic display. Yet color monitors have dark backgrounds instead of the more familiar white, and video graphics with large amounts of white can "bloom" and irritate the eye. The problem need not lie with the computer—software engineers with no training in cartography and little sense of graphic design have been highly successful in developing online mapping applications. With no guidance and poorly chosen standard symbols, users of mapping software can be as accident-prone as inexperienced drivers. If you see one coming, look out!

Maps That Advertise

What do advertising and cartography have in common? Without doubt the best answer is their shared need to communicate a limited version of the truth. An advertisement must create an image that's appealing and a map must present an image that's clear, but neither can meet its goal by telling or showing everything. In promoting a favorable comparison with similar products, differentiating a product from its competitors, or flattering a corporate image, an ad must suppress or play down the presence of salt and saturated fat, a poor frequency-of-repair record, or convictions for violating antitrust, fair-employment, and environmental regulations. Likewise, the map must omit details that would confuse or distract.

When the product or service involves location or place, the ad—whether in print or electronic form—often includes a map, sometimes prominently. Two mildly Machiavellian motives account for the use of maps as central elements in some advertising campaigns and as important props in others. First, as art directors and marketing specialists have discovered, the map's need to avoid graphic interference can serve the advertiser's need to suppress and exaggerate. Indeed, maps that advertise tend to be more generalized than graphic clarity demands. Second, ads must attract attention, and maps are proven attention-getters. In advertising, maps that decorate seem at least as common as maps that inform.

This chapter examines cartographic distortion in commercial advertising. It looks first at transportation ads, which frequently employ maps to exaggerate the quality of service. Occasionally the distortion is deliberately overdone to make the map a graphic pun that uses artistic flair to make the advertiser's point about convenience or improved service. The second section shows how a map touting one or a few locations can convey an image of convenience or exclusiveness. The final section treats maps that promote chain stores and franchise businesses by equating numerousness with success and quality. Simple scenarios illustrate how advertisers and creative agencies exploit maps as marketing tools. The contrived levity of the examples discussed reflects real advertisers' reluctance to submit to public criticism as well as the light-hearted, half-serious attitude of many ad maps.

Transport Ads: Gentle Lines and Well-Connected Cities

It is 1875. As president of the soon-to-be-completed Helter, Skelter, and Northern Railway (HS&N), you need to advertise your route both in a timetable and in the shippers' bible, the *Official Guide of the Railways*. Your engineering department's small-scale map (fig. 6.1) seems not quite suitable. No point in drawing attention to your principal competitor, the Helter, Skelter, and Yon, whose more direct route actually reaches downtown Skelter rather than terminating in West Skelter, 3 miles from the Skelter business district. Nor would you want to publicize the HS&N's tortuous trek up the floodplain of Catfish Creek to the wishful city of Bogsville, which willingly bought construction bonds to attract a railroad. And since you seek further capital, the map might also add credence to the still dubious promise "and Northern" in the company name. Further, the overall shape and geometry of the engineer's map seems inappropriate: you want the HS&N clearly in the center, and you need room to show the names of as many cities, towns, and hamlets as you dare claim for your sparsely populated right-of-way.

You explain your needs to a local artist with experience working for railroads, who three days later proudly delivers the map in figure 6.2. Amazed, pleased, and grateful, you first stare at the map for a minute and wonder whether it shows the correct railroad. Its dominant visual feature is a nearly straight line connecting Helter

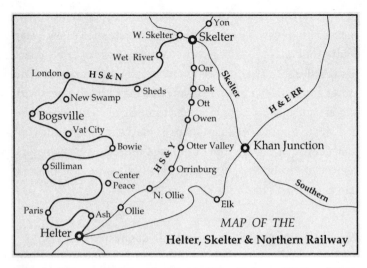

Figure 6.1. Engineering department's map of the Helter, Skelter, and Northern Railway.

with Skelter. Having captured all towns and hamlets close enough to warrant a crossroads sign or unoccupied shelter, your cartographic fiction shrewdly suppresses the label "West Skelter" and suggests a direct link with other rail lines in Skelter. (If naive viewers want to assume a common connection downtown, well, that's their inference.) And a nearly-as-prominent dashed extension running off the map to the north suggests imminent new construction and vast possibilities for investment and profit. In contrast, a thin, graphically weak line portrays your principal competitor, the Helter, Skelter, and Yon Railroad, as not only offering a more devious route between Helter and Skelter but also nearly bypassing Yon. Thanks to cartographic license, the Helter, Skelter, and Northern Railway has become an attractive option for distant shippers and distant investors.

Time flies, and it is now the early twenty-first century. In a new reincarnation, you are president of the recently formed Upward Airlines. And once again you need a map for media advertising. Yet the cartographic challenge has changed: instead

Figure 6.2. Advertising and timetable map of the Helter, Skelter, and Northern Railway.

of de-emphasizing a sinuous route and ignoring a single parallel competitor, you now need to draw attention to the number of cities served and the overall integration of the Upward Airlines system. Like other national airlines, Upward operates a hub-and-spoke system, with planes converging several times a day on each of its two hubs, where passengers get off, hike to another gate, and board a second aircraft.

Planes fly fairly direct routes, as everyone knows, so your modern-day advertising agency freely sacrifices straight lines and their impression of geographic directness for the dramatic, rather busy-looking map in figure 6.3. Cartographic license lies largely in the map's suggestion that service is equally frequent along all the spokes. In reality, Bismarck, North Dakota, has only one plane in and out each day, and then only in early afternoon, which is inconvenient for business travelers who don't relish spending an extra night in Bismarck. And if fewer than 20 percent of their seats

Upward Airlines — Reach for the clouds

Figure 6.3. Advertising map emphasizing Upward Airlines' service area and connections.

are sold, flights from Shreveport, Louisiana, to St. Louis are often canceled because of bad weather or "mechanical problems."

Another expediently misleading generalization is the map's suggestion that connections at Upward's hubs are convenient. In addition to omitting the reality of frequent, frustrating delays on taxiways and in crowded boarding areas as well as less frequent but more frustrating overnight stays because of missed late-evening connections, the hub symbol tells the prospective passenger nothing about the likelihood of a quarter-mile walk with infant, garment bag, or other "carry-on items" from gate A-11 to gate E-12. Formed by the merger of Westward Airlines and Northeastward Airlines, Upward Airlines must make do with its predecessors' gates at opposite ends of the terminal. Because this situation might trigger anxiety about missed connections, a floor plan is unlikely to appear in Upward's in-flight magazine.

Advertising agencies serving airlines can have great fun with maps if the client has a sense of humor. In addition to decorating maps with pictograms of impressive skyscrapers, museums, golfers, vacationers in bathing suits, and other symbols of culture or

leisure, graphic designers can create a variety of cartographic puns
by manipulating maplike images, like the converging routes that
suggest a duo of octopuses (fig. 6.3).

Sponsorship and Visibility: Ad Maps with a Single-Place Focus

In an ad promoting a store, resort, or other business, a map might
not only offer travel directions but also stimulate demand. For many
goods and services, especially those readily and conveniently available from online vendors, the trip itself is an important consider-
ation in the purchase. If bad roads, walking through or parking in
an unsafe neighborhood, or heavy traffic would make the trip an
ordeal, the buyer might reasonably consider travel an added cost
and simply order the product from her laptop. Thus a map promoting a retail outlet needs not only to suggest straightforward routes
for getting there but also to present an image of convenient accessibility. And if an attractive image requires distorted distances and
directions, the advertising map will indeed distort. Accuracy and
precision, after all, are seldom prime goals in advertising.

Accessibility is particularly important for products needed in a
hurry. And few customers feel as much duress as the do-it-yourself
plumber confronting a sudden shower in the basement or a running
toilet that tinkering has only made worse. Thus when Rudy Swenson
stops by your newly opened creative agency to discuss a campaign
for his plumbing-supplies house, you immediately impress him by
suggesting a specially designed map to feature on his website, which
already conveys the message that plumbing repairs are easy if you
have the right parts and tools. Like other shoppers, after all, do-it-
yourself plumbers are conditioned to respond to emergencies by
looking online for a well-stocked parts dealer.

Rudy's business is ripe for a tailor-made ad map, beyond the
generic street map that might appear in the search engine's results.
None of his few competitors in town are on the same street, so your
map need not give circuitous directions to avoid a business rival.
Although his store is in the low-rent section of the city, the two

Rudy's Plumbing Supplies: as close as you are

Figure 6.4. The map used on the website for Rudy's Plumbing Supplies does more than give directions.

streets intersecting at the corner extend far enough out of town to avoid the immediate area's reputation for vice and mayhem. (A trip to this area might even appeal to the do-it-yourself plumber's penchant for living on the edge.) Although you briefly consider sprinkling some imaginative travel-time estimates across the map, you decide instead to invoke the advertising cartographer's ever-reliable ploy—place-name dropping. Your finished design (fig. 6.4) shamelessly mentions a host of neighboring towns, some of which have their own plumbing-parts dealers. Yet the thought of customers traveling 30 miles over State Route 10 from Canton and a good 45 miles over tortuous State Route 19 from Cynwyd pleases Rudy as much as it impresses local homeowners. And right near the center of the map is a drawing of Rudy's building, its perspective seeming to justify your wild and willful distortion of the region's geography. (What newcomer would guess that East Hills is actually closer to Rudy's than Westvale? But rather than take offense, residents of both places are tickled to see their towns mentioned. In fact, only persons whose towns aren't shown are likely to complain.)

Your second client, Karen Torricelli, is less concerned about convenient accessibility or online competitors. Her business offers not a quick shopping trip but a recreational retreat: people go to Karen's Bowling Camp to get away from it all, to commune with nature, to hike and swim and fish, and to bowl. Although the map that meets Karen's needs might also benefit from the geometric distortion used on Rudy's map, it requires a different kind of geographic name-dropping.

Your map (fig. 6.5) offers an attractive solution for the Bowling Camp's brochure. Its prominent position at the top of the map fits with the local sense of its being "up north." You tout its accessibility via "scenic" Interstate 32 and imply that it is an attractive vacation spot for happy campers from as far away as Cleveland, Pittsburgh, and Knoxville. And you reproduce a miniature image of the highway exit sign to reinforce Karen's association with Lakeport and the Lake Walleye resorts. Of course, the only other resort you show is the comparatively upscale Kelly's Yacht Club, which also serves as a convenient landmark. Moreover, the town of Lakeport is "historic Lakeport." You give the area a cheery image with tiny pictorial symbols to mark an abandoned nineteenth-century military base and to suggest pleasant times to be had fishing in and sailing on Lake Walleye. Other tourist spots mentioned include the boyhood home of sage and renowned debtor Owen Moore, the unique, world-famous Stain Museum, and that delightful monument to Plastic America, Tex and Edna Boyle's Curio Emporium. Not exactly Lake Tahoe, central New Hampshire, or the Wisconsin Dells, the Lake Walleye area has a charm perhaps best appreciated by Tex and Edna's customers and Karen's guests. And your map captures this ambiance to perfection.

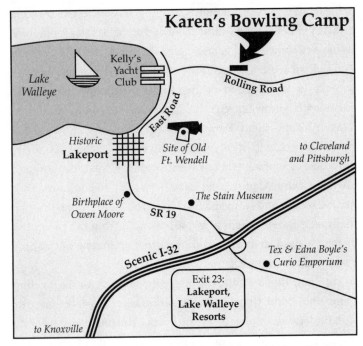

Figure 6.5. The map giving directions to Karen's Bowling Camp also shows the viewer other attractions in the area.

You push the use of pictorial icons further by convincing local civic organizations to issue walking-tour maps that highlight local landmarks and sponsoring businesses. In some towns the maps are distributed free to local diners that use them as place mats; in others, they're offered as tourist flyers in take-me racks in motel lobbies and highway rest stops. Businesses pay to get on the map; those that don't pony up have their spots filled with trees, mundane structures, or children at play.

Spurred by the successful use of maps in your advertising campaigns for Rudy and Karen, you decide to reach out to potential clients elsewhere in the nation. Having seen a number of catchy ads using humorously distorted maps, you decide to make the cartographic pun a hallmark of your agency. After a few abortive attempts to capitalize on the hourglass shape of Scranton, Pennsyl-

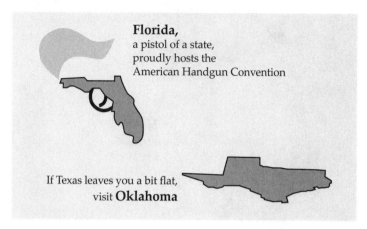

Figure 6.6. Cartographic puns promoting a handgun convention in Florida (above) and in-state tourism among Oklahomans (below).

vania ("For the time of your life visit Scranton!"), and the Miami–Fort Lauderdale, Florida, area's resemblance to an alligator ("Snap up your lot or condo today!"), you decide to work only with geographic shapes familiar to large numbers of nonlocal people. So after much thought and anguish you develop the kernels of two ideas, one for the Florida Convention Bureau and the other for the Oklahoma Convention Bureau. As figure 6.6 shows, the former is for a somewhat targeted—er, highly specific—campaign, which would not run in the general media, and the other is designed for obvious reasons to run only in Oklahoma, not in Texas. Hmmm. What could you do with New Jersey?

Numerousness and Territoriality: Ad Maps
Touting Success and Convenience

Not all of your out-of-town clients have been as unappreciative as the Scranton and Miami tourist bureaus. Indeed, one of your biggest recent successes was a map-supported campaign developed for Holly's Hotdish Heaven, a regional restaurant chain with headquarters in Minneapolis. Holly's specializes in a Minnesota

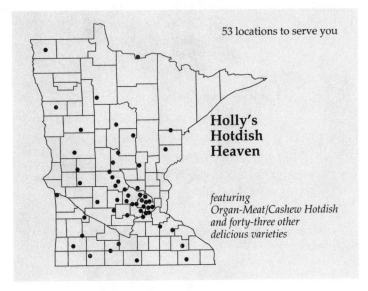

53 locations to serve you

**Holly's
Hotdish
Heaven**

*featuring
Organ-Meat/Cashew Hotdish
and forty-three other
delicious varieties*

Figure 6.7. Ad map proclaiming the widespread local acceptance of a regional restaurant chain.

specialty, hotdish. In fact, all fifty-three restaurants are in Minnesota. (Unsuccessful attempts to establish outlets in Iowa, South Dakota, and Wisconsin suggest that the preference, or tolerance, for hotdish is highly regional.)

Anyway, your map (fig. 6.7) works nicely for Holly's, which plans ten new outlets over the next five years, all in Minnesota. The map exploits a simple strategy common to advertising firms with multiple outlets or many far-flung clients: numerousness indicates success, and success indicates a superior product. The hungry Minnesotan has only to look at the map to see that Holly's has been successful. The dots indicate acceptance by discriminating palates throughout the state, but especially in the Twin Cities area toward the southeast, where more people live. Indeed, closing the three experimental restaurants in neighboring states meant that this single-state base map can not only show a fuller, more successful-looking packing of the dots but also promote a stronger identifica-

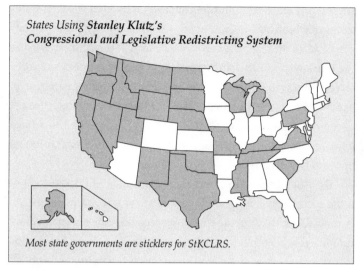

States Using **Stanley Klutz's**
Congressional and Legislative Redistricting System

Most state governments are sticklers for StKCLRS.

Figure 6.8. An ad map equating numerousness and land area with product quality exploits sales to larger states in the West.

tion with the state and its people's pride in their traditions. If Holly's becomes much more successful, you'll have to use a smaller dot.

Another recent cartographic success is your map for Stanley Klutz Associates, a software developer and vendor active in geographic information systems (GIS). One of Klutz's best-selling products is a GIS designed to help states reapportion their congressional and legislative districts while helping partisan politicians craft safe districts or box their opponents into districts where they are little known and out of step with most voters. Klutz redistricting software is widely recognized as providing an acceptable compromise that somehow meets the needs of powerful politicians now in office and satisfies the Supreme Court's notorious one person/ one vote decision.

Eager to increase its market penetration, Klutz called after hearing of your spectacular success in the Minnesota hotdish campaign. Its marketing manager was particularly impressed by the map you created for them (fig. 6.8), which most effectively exploits Klutz's success among larger but often sparsely inhabited western states (some of which have only one congressman anyway). Indeed, your map creates the impression that well over half the country is using the Klutz system, when in fact twenty-six of fifty states use another vendor's system or the time-honored tradition of paper map, pencil, eraser, and the smoke-filled back room. Of course, if most of their success had been in the eastern states, a map could have been linked to the slogan "Get with it, America—there's still lots of room for a Klutz!"

Like other artwork in commercial print and online advertising, maps can be clever and catchy as well as contrived and deceptive. In most cases, though, the consumer recognizes the map as a playful put-on and appreciates being in on the joke. The ad map thus is further evidence of the map's enormous flexibility and appeal. Less benign, of course, are maps hawking remote building lots and doubtful mineral rights as well as many other maps that attempt to advocate or seduce. The next two chapters treat the use of maps as tools for real-estate developers and political propagandists.

CHAPTER 7

Development Maps
(Or, How to Seduce
the Town Board)

Without maps, urban and regional planning would be chaotic. Detailed maps describe the relative sizes, shapes, and spacing of a plan's components and suggest how well they interrelate. Maps of a planned shopping plaza, for instance, would show the overall shape of the project, the sizes and general layouts of individual stores and public spaces, the sizes and locations of parking areas, landscaping, driveways, and entrances from a public road. Public officials use these maps to assess the plaza's impact on nearby neighborhoods, traffic, and established businesses. Overlaying the plan onto topographic and soils maps reveals the project's likely effects on wildlife, wetlands, and streamflow, as well as potential difficulties with sinkholes, unstable soils, or a high water table.

Even with maps, many cynics would argue, urban and regional planning is chaotic. As an inherently selective view of reality, the map often becomes a weapon in adversarial negotiations between developers and the local planning board. After all, the developer might have millions of dollars invested, whereas nearby residents probably don't want a new retail center in their backyards, and planning boards often reflect local fears and biases. The developer's maps thus attempt to impress residents of more distant neighborhoods with the project's elegance and convenience and to demonstrate that it will harm neither wildlife nor property values. The residents' maps, if they are used at all, will focus on habitat destruction, traffic congestion, visual blight, noise, and trash. Because the developer has deeper pockets, the opposition's maps will not look as nice as the developer's graphics, which suppress dumpsters, litter, and abandoned cars but optimistically portray skinny saplings as mature shade trees.

This chapter examines the role of maps in city and regional planning, especially as a tool of persuasion. It begins with a concise introduction to the part mapping plays in zoning and environmental protection, examines how developers might manipulate maps to enhance their cases, and concludes with a short example of how an overtaxed homeowner can use maps to argue for a reduced real-property assessment. It complements chapter 13, which focuses on prohibitive cartography, including zoning maps.

Zoning, Environmental Protection, and Maps

Zoning refers to the legal process municipalities use to control land use and land subdivision for development. Although legislation controlling the operation and powers of local planning and zoning boards varies from state to state, zoning laws typically regulate the height, size, character, and function of buildings; the minimum size of a building lot and the location of the building on the lot; and associated improvements such as outbuildings, driveways, and parking lots. This system of laws, hearings, and enforcement procedures, which gained wide acceptance early in the twentieth century, is essential for effective urban planning.

Community planning boards commonly work with three principal maps: (1) an *official map* to show existing rights-of-way, administrative boundaries, parks and other public lands, and drainage systems; (2) a *master plan* to indicate how the area should look after several decades of orderly development; and (3) a *zoning map* to show current restrictions on land use. Figure 7.1 (plate 5) shows the legend and a portion of a typical zoning map for a town far enough from a major city to still have working farms. The various categories reflect differences in the density of people and dwelling units, whether the area is generally open to the public and offers a pleasant view, and the likelihood that noise, litter, or other nuisances might affect adjoining areas. For each zoning category

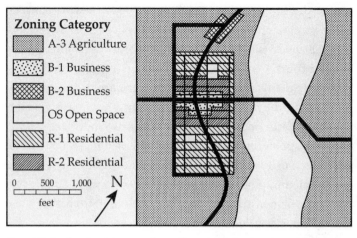

Figure 7.1. Portion of a zoning map of a small community in a largely rural area.

listed in the key, a set of restrictions applies to define precisely what kinds of structures and activities are permitted. In an R-1 residential district, for instance, a lot must have a minimum size of 5,000 square feet and a minimum width of 45 feet and may contain a single-family home that covers no more than 50 percent of the lot and is set back at least 20 feet from the front of the lot and 5 feet from each side.

Zoning boards spend considerable time hearing requests for *variances* so that property owners can do things that otherwise would violate the zoning ordinance. For example, a homeowner might want to add a room or a garage that would encroach within 10 feet of the property line. Or a business might want to purchase and raze a neighboring structure, grade the land, and make a parking lot. Zoning laws recognize that any improvement affects far more than just the ground underneath, but these laws also allow exceptions that meet community approval and are not too obnoxious. The zoning board must examine plans describing proposed alterations, then solicit and listen to the opinions of neighbors and other interested citizens. A board can grant or deny variances as proposed, grant modified variances with specific restrictions added, or grant temporary variances. The applicants or their neighbors can appeal the board's decision to a board of zoning appeals or to the court.

Major modifications of the zoning map are treated as modifications of the master plan and require a hearing before the planning board. Housing developments, new subdivisions, retail centers, large stores, and other types of development that add or alter streets or affect municipal services such as water and sewage require planning-board approval. The developer of a shopping plaza must demonstrate, for instance, that runoff from its parking lots will not cause flooding and that existing roads can carry the increased traffic. Sometimes the developer must work with the county or state highway department in planning new access roads, to be paid for directly or through an "impact tax." The developer of a subdivision must provide a map describing lot boundaries and showing roads and utility lines.

Planning boards are very concerned with housing density—that is, the number of dwellings per acre. Many municipalities specify a minimum lot size as large as 3 or 5 acres, ostensibly to "preserve the environment," but often to exclude lower-income families unable to afford such expensive housing. But sometimes a planning board will approve a development of town houses or clusters of four or more dwellings with sufficient open space to meet a minimum *average* lot-size requirement. Maps are a convenient format for describing the developer's proposal and are an essential part of any subdivision hearing.

In the 1960s, growing concern about environmental degradation led to a substantial increase in environmental regulations and in the number and size of public agencies that monitor compliance. Although practices vary from place to place, state or municipal environmental-quality review boards prepare inventories of wetlands and other sensitive wildlife habitats, vegetation, surface water, groundwater, soils, slope, geology, and historic sites. Commonly compiled using soil-survey maps, aerial photography, or

existing topographic maps, these *environmental resource invento-*
ries are used to assess the likely adverse effects of highways, retail
centers, residential tracts, landfills, industrial plants, and other
types of development. If the inventory is accurate, town or county
planners can quickly tell whether a proposed project is likely to
affect a fragile wetland habitat or contribute significantly to the
extinction of a rare plant.

Large projects, whether public or private, commonly require an
environmental impact statement (EIS). The developer usually must
supplement information in the environmental resource inventory
with field measurements compiled by an environmental consulting firm of civil engineers, landscape architects, geologists, and
biologists. The list of possible areas of impact the EIS might address
includes air pollution, water pollution, public health, hydrology
(such as the possibility of increased flooding downstream), ero-
sion, geologic hazards such as earthquakes and landslides, flora
(especially rare plants), fauna (especially endangered species),
wildlife in general, the aesthetic and scenic values of both the natural landscape and the built environment, solid waste, noise, the
social environment, economic conditions, recreation, public utilities, transportation, and the risk of accidents. An EIS must identify
the types and severity of plausible impacts, the areas affected, and
alternative strategies with a lesser impact.

Because preparing an EIS is costly and time-consuming, in
many cases a shorter, less comprehensive *environmental assessment*
can demonstrate that the impact will be minimal, that the project
will comply with environmental regulations, and that an EIS is not
necessary. Environmental-resource inventories can enable local or
state environmental-review boards to evaluate the environmental
assessment and either approve the project or require a full EIS.

Maps are an important part of an EIS or environmental assessment. Environmental scientists commonly transfer all map information to a common base for ready comparison. Sometimes a computerized geographic information system is used to store the data
and generate final plots. Detailed, oversize maps might accompany

the EIS in an appendix, to supplement smaller-scale, more generalized maps in the body of the report, which might be made available online. (How many readers bother to compare the large-scale maps with their small-scale generalizations, positioned much closer to the analytical and persuasive parts of the report?) Potentially significant sources of error are the transfer of information from the source map to the common base and the generalization of these small-scale maps. (When the alternatives are equally valid, can the consultant resist the temptation to draw the line that favors the client's case?) Additional problems arise when boundaries and other data are transferred from unrectified aerial photographs (see chap. 3)—in hilly areas these lines are probably not accurate unless the mapmaker adjusted for relief displacement. Persons opposing a project might well begin by taking the EIS into the field and checking its supporting maps against the features portrayed.

Sometimes the data simply cannot reflect what the developer or the compliance agency would like them to show. As examples, floodplains defined locally by a single elevation contour tend to include either too much or too little of the real floodplain, and soil-survey maps might not reliably reflect the depth to bedrock, an important indicator of where septic tanks and leach fields are unsuitable. When using soils maps to compile maps for an environmental assessment, the developer should resist the temptation, illustrated in figure 7.2, to alter categories, replace a technical definition with a questionable interpretation, or "inadvertently" omit small parcels of land whose presence might disqualify a larger tract. Both the developer and the review panel should be aware that soils maps are based on a soil scientist's interpretation of the land surface and a limited number of subsurface core samples. Moreover, soils maps generally do not show patches smaller than the head of a pencil, which at 1:20,000 would cover several generous building lots. When the developer's consulting engineer carries out a special field survey, a map should show the actual locations of the subsurface core samples to allow the viewer to verify that the map is neither much more nor much less generalized than the data it represents.

Depth to bedrock:
21 – 48 inches

Depth to high water
table: 2 – 4 feet

Soils 48 inches
deep

High water table
4 feet deep

Figure 7.2. Creative generalization and interpretation of the soils map and categories on the left can yield the more favorable cartographic representation on the right.

Zoning cases and environmental-quality reviews often move from the administrative hearing to the courtroom. When deliberations escalate to the judicial level, maps become important as exhibits, together with aerial photographs, drawings and architect's renderings, scale models, ground-level photographs, and videos. The courtroom audience usually is more sophisticated and less harried than the local planning or zoning board, and both parties commonly employ technical experts to testify as well as to advise in cross-examination. Ploys that might have impressed a volunteer group of businesspeople, farmers, teachers, and retirees can fail or backfire, and exhibits must be easily defended as well as convincing and persuasive. Opposing sides often differ principally in their interpretation of identical exhibits.

Maps are indispensable exhibits in land-use cases, which cannot be prepared and presented without them. A minimal presentation would include the master plan, a detailed zoning map of the affected area, and an enlarged aerial photograph or two. Attorneys and witnesses often use a marking pen or tape to identify locations, and they require multiple copies of most exhibits so that these marked-up materials can become part of the record. The attorney carefully marks each exhibit for introduction as evidence, identifies its source, calls an appropriate witness to explain it, and reviews in

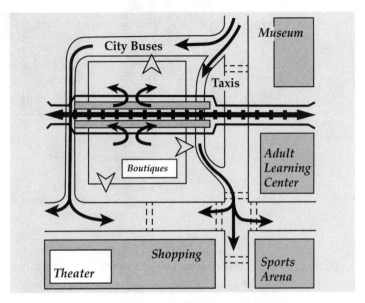

Figure 7.3. A concept diagram for a proposed downtown transportation center with rail, bus, and taxi service.

advance the scope of the witness's testimony and plausible counterarguments by the opponent.

Although topographic maps, air photos, and zoning maps tolerate no embellishment, the developer has considerable license in the design and content of site plans and architect's renderings. A particularly intriguing and forceful graphic is the *concept diagram*, a schematic, somewhat stylized map intended to demonstrate the general layout and functional relationship of a plan's main elements. The example in figure 7.3, which illustrates the interchange concept for a downtown transportation center, shows how the developer or planner uses lines to subdivide space, highlight patterns of movement, and suggest revitalization of the central city. Concept diagrams have a compelling, mysterious attraction and can be highly persuasive when explained by an enthusiastic architect. These maps encourage the viewer to want to see the plan work, not to wonder whether it will work. Once the concept diagram has convinced the audience that a project is functional and feasible, the presenter can introduce three-dimensional models, sketches, and other persuasive renderings to show how the finished project should look. Like most utopian views of the future, these maps and pictures feign realism with selective detail.

Tricks for Polishing the Cartographic Image

Over time, developers have identified a number of ways to play down the adverse impact of a proposed project and enhance its visual appearance and presumed benefits. These tricks work best before a town zoning or planning board and are least effective in state or federal court.

1. *Be shrewdly selective.* Don't show what you'd rather they not see. Omit potentially embarrassing features that might evoke unpleasant images of litter, congestion, and noise. Omit dumpsters and other trash containers, traffic signs, loitering teenagers, and trucks. In sketches and scale models either omit people and cars altogether or show only smartly dressed people and late-model, fuel-efficient cars. Never admit the possibility of dying shrubbery, trampled turf, or anything remotely suggesting a plume of smoke. Above all, keep the image clean and sufficiently generalized so that such omissions don't appear unnatural.

2. *Frame strategically.* Avoid unfavorable juxtaposition, and crop the maps and sketches to forestall fears of illness or diminished property values. If a neighboring site is unattractive or likely to be unfavorably affected, leave it out. If the proposed development adjoins a park or another attractive site, leave it in. If a neighboring property might also be improved, include it but show it too as newly developed. Never show a school or homes bordering a proposed landfill, solid waste treatment facility (the euphemism for municipal incinerator), or electric power lines.

3. *Accentuate the positive.* Choose favorable data and supportive

themes for maps. If, for instance, a proposed landfill will have a high fence or unobtrusive entrance, by all means show it. If a new shopping plaza would displace an existing eyesore, a set of "before" and "after" maps is useful. Favorable interpretations of data or source maps also help.

4. *If caught, have a story ready.* Software errors, an intern's unfortunate use of the wrong labels, or the accidental substitution of an earlier version of the map make plausible excuses.

5. *Minimize the negative.* If you can't eliminate them entirely, at least don't emphasize features you'd rather have ignored. Note that the train station in figure 7.3 doesn't call attention to exhaust from idling buses and taxis.

6. *Dazzle with detail.* After all, a detailed map is a technically accurate map, right? Details are useful distractions.

7. *Persuade with pap.* Try highly simplistic maps, or maps with fire hydrants, mailboxes, and any other irrelevant minutiae that might camouflage potentially embarrassing details.

8. *Distract with aerial photographs and historical maps.* These make great conversation pieces and are excellent distractions for people eager to exclaim, "Hey, there's my house!"

9. *Generalize creatively.* Filter or enhance details to prove your point. A little selective omission or massaging of contours, soils boundaries, or even property lines might well pass for cartographic license.

10. *Enchant with elegance.* And don't forget the architect's cartographic friend, the tree stamp, or its electronic counterpart. As figure 7.4 demonstrates, symbolic trees can convert a mundane proposal into a pleasant neighborhood asset, and the more of these hypothetical trees, the better. After all, it takes much less time and effort to add treelike symbols onto the map than to plant the real thing. And in twenty years those anemic saplings you will plant might even resemble the healthy shade trees in the picture.

11. *When all else fails, try bribery.* Not under-the-table monetary payoffs, of course, but such institutional bribery as decent-

No
trees

Few
trees

Many
trees

Figure 7.4. Tree symbols add visual appeal to an otherwise barren developer's plan.

paying jobs for the unemployed, good profits and well-paying jobs for contractors and construction workers, a larger tax base or "payments in lieu of taxes" for local government, prestige, and promises of amenities elsewhere in the community for teenagers, young families, and senior citizens. Or try another area, where citizens and their representatives are less aware of graphic trickery.

The Assessment Review

Like words and numbers, maps are anybody's weapon, and they can also help the homeowner appeal an unfairly high tax assessment. But whether the principles that follow are useful will depend on how a given area estimates real-estate values.

In most areas, the municipality computes the yearly tax on

"real property" by multiplying the assessed value of a building and lot by the established local tax rate. The total value of land and buildings in the area and its estimated expenses for schools, government operations, welfare, and debt service determine the tax rate. Each parcel's assessed value is someone's guess of what the property is worth. In areas with "full-value assessment," this guess is an estimate of the current fair market value of the property. In other areas the assessed value is supposed to be a fixed percentage of the fair market value. But other factors, some political, often influence the guessing.

Assessment practices vary widely. Some jurisdictions rigorously apply a set of guidelines, or formulas, based on measurements such as lot size, area of the dwelling, and number of bedrooms and bathrooms. Other jurisdictions filter these objective criteria through the subjective judgment of an assessor or assessment board. Since location is an important part of a property's worth, assessors not slavishly governed by formulas will consider scenic values and other less tangible amenities as well as recent sale prices of nearby properties.

Because housing values change over time in response to inflation, new amenities and nuisances, and shifting perceptions of which neighborhoods are good and which aren't, assessors usually spend more time reassessing old properties than evaluating new ones. Some areas reassess yearly or on a regular cycle of every two, three, or four years, whereas others reassess more randomly, according to the judgment of the assessor.

A common practice is to reassess only when a property is sold or altered by a major structural change such as adding a room or a fireplace. In these areas assessments tend to favor long-term residents who bought property many years ago and have made no major improvements. Widespread reassessment would tend to hurt these people, especially if their neighborhoods have not deteriorated. Because long-term residents tend to vote regularly as well as oppose new assessment practices that might be fairer, a directly elected assessor or one appointed by the town board is likely to reflect their interests. The result is an assessment scheme aptly called "soak the newcomer" or "Welcome, stranger."

But assessments, like zoning decisions, can be appealed. In larger municipalities, a formal appeal might even require the services of an attorney. Towns and villages, more likely to reassess mostly newcomers, have a more relaxed approach—an annual "grievance day" that might even be a week or a month long. Residents come before the board to "grieve," that is, to challenge their assessments. Yet a successful appeal will depend not on griping about the assessor's incompetence or the cost and quality of municipal services but on demonstrating convincingly that the proposed assessment clearly is out of line. In short, a homeowner needs to do some research.

The assessor's office is a good place to begin. In many jurisdictions, this information is available online. To show that you have been overassessed, you might compare your assessment with those of other properties in the neighborhood, especially similar houses. Come prepared with the addresses of properties whose records you want to examine. Workers in the assessor's office will help you locate a description of each property and its assessment history. These records show noteworthy improvements and special features and list the sizes of the lot, the living area, and perhaps individual rooms. If your assessment is not radically different from those of similar dwellings, you probably have no case—but count yourself fortunate to have been treated fairly from the outset. Yet if your assessment is well above those of similar properties nearby, an appeal could save you considerable money, especially if you stay for several years.

You can present your appeal using two types of evidence: facts showing that the properties you think are similar are indeed so, and comparative figures demonstrating your new assessment is too high. (Should you find that one of these "comparables" is also taxed unfairly, you might ask your neighbor for support in a joint appeal.) Three kinds of exhibits can help establish similarity: a map of the neighborhood, a poster with photographs and street

DEVELOPMENT MAPS (OR, HOW TO SEDUCE THE TOWN BOARD)

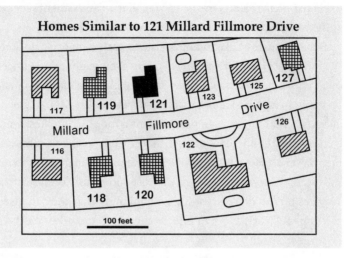

Figure 7.5. A map showing your home and nearby houses used for comparison.

addresses, and tables or charts comparing your house with the others for type of construction (wood frame, brick, stone, etc.), number of rooms, area of living space, year of construction, and lot size. Figure 7.5, a typical neighborhood map compiled by tracing street, lot, and foundation lines from the assessor's maps, can show both similarity and proximity. As an able propagandist, you naturally choose a title such as "Similar Properties in the Neighborhood" to reinforce your point.

Having convinced the board that these houses are similar to your own, you now present a map dramatizing disparities in assessment. As in figure 7.6, the mean assessment for the group of similar properties (excluding yours) can be a useful basis for graphic comparison as well as a subtle hint of a fair assessment for your own parcel. As a referent, this mean assessment also allows you to use symbols that focus on *differences* in assessments, so that comparative bar symbols more effectively represent unfairness. If you can win the argument for similarity, this map can win your case for a lower assessment.

If your proposed assessment exceeds that of any of your neigh-

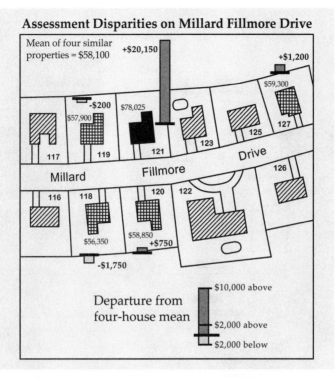

Figure 7.6. A map showing disparities in assessed value between your house and similar properties in the neighborhood.

bors, you might want to lead by comparing your property with the parcel having the next highest assessment. For this presentation use a simple map showing lot locations, addresses, and assessments. As figure 7.7 illustrates, the map's symbols and labels should show board members that you know about assessments in your neighborhood and that you have a legitimate grievance. Then introduce photographs and tables showing that your house is much more modest in appearance, size, and amenities than the one with the second-highest assessment. If the assessor has flagrantly been playing "soak the newcomer," your exhibits will send a clear message that you are fully prepared to embarrass the board unless they make an appropriate adjustment. If you use tact, most of the time they will.

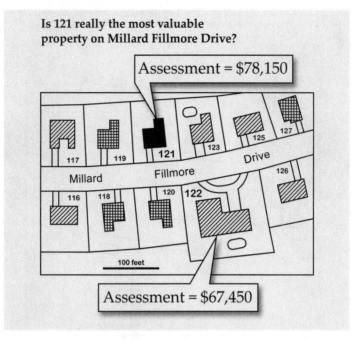

Figure 7.7. A map to demonstrate that your proposed assessment ought not to be the highest in the neighborhood.

Although the tone of this chapter is cynical, the intent is to make you skeptical about how some people use maps, not cynical about maps in general. Understanding cartographic manipulation is important to being an informed citizen who is able to evaluate a wide range of proposals for altering the landscape and the environment. In viewing maps it is essential to remember that a particular view of reality (or a future reality) is not the only view and is not necessarily a good approximation of truth.

Maps for Political Propaganda

A good propagandist knows how to shape opinion by manipulating maps. Political persuasion often concerns territorial claims, nationalities, national pride, borders, strategic positions, conquests, attacks, troop movements, defenses, spheres of influence, regional inequality, and other geographic phenomena conveniently portrayed cartographically. The propagandist molds the map's message by emphasizing supporting features, suppressing contradictory information, and choosing provocative, dramatic symbols. People trust maps, and intriguing maps attract the eye as well as connote authority. Naive citizens willingly accept as truth maps that are based on a biased and sometimes fraudulent selection of facts.

Although all three manipulate opinion, the propagandist's goals differ from those of the advertiser and the real-estate developer. Both the advertiser and the political propagandist attempt to generate demand, but the advertiser sells a product or service, not an ideology. Both the advertiser and the propagandist attempt to lower public resistance or improve a vague or tarnished image, but the advertiser's objectives are commercial and financial, whereas the propagandist's are diplomatic and military. Both the real-estate developer and the political propagandist seek approval or permission, but the developer is concerned with a much smaller territory, often uninhabited, and seldom acts unilaterally without official sanction. Although both the real-estate developer and the propagandist face opponents, the developer usually confronts groups of neighboring property owners, environmentalists, or historic preservationists, whereas the propagandist commonly confronts a vocal ethnic minority, another country, an alliance of countries, an opposing ideology, or a widely accepted standard of right and wrong. Because propaganda maps are more likely to be global or continental rather than local, the political propagandist has a greater opportunity than either the advertiser or the real-estate developer to distort reality by manipulating the projection and framing of the map.

This chapter explores the map's varied and versatile role as an instrument of political propaganda. Its first section examines how maps function as political icons—symbols of power, authority, and national unity. Next the chapter looks at how map projections can inflate or diminish the area and relative importance of countries and regions, and how a map projection can itself become a rallying point for cartographically oppressed regions. A third section examines the manipulations of Nazi propagandists, who used maps to justify German expansion before World War II and to try to keep America neutral. A final section focuses on a few favorite symbols of the cartographic propagandist: the arrow, the bomb, the circle, and place-names.

Cartographic Icons Big and Small:
Maps as Symbols of Power and Nationhood

The map is the perfect symbol of the state. In past centuries, if your grand duchy or tribal area seemed tired, run-down, and frayed at the edges, you could simply take a sheet of paper, plot some cities, roads, and physical features, draw a heavy, distinct boundary around as much territory as you dared claim, color it in, add a name—perhaps reinforced with the impressive prefix "Republic of"—and presto: you were now the leader of a new sovereign, autonomous country. Should anyone doubt it, you merely had to point to the map. Not only was your new state on paper, it was on a map, so it had to be real. Today's would-be sovereign could do the same with graphics software.

Plate 1. Linear cartogram of the Washington, DC, Metro system.

Point Symbols	Line Symbols	Area Symbols

Plate 2. For area symbols in particular, hue is often more forceful than the other five principal visual variables (fig. 2.11).

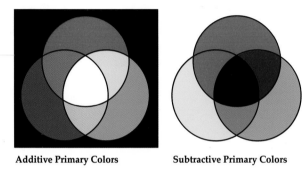

Additive Primary Colors   **Subtractive Primary Colors**

Plate 3. Primary colors yield other hues when combined as either beams of light (left) or patches of dye (right).

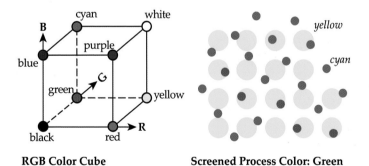

RGB Color Cube   **Screened Process Color: Green**

Plate 4. RGB color cube (left) and greatly enlarged representation of overprinted screens of colored dots used in process printing to produce green (right).

Plate 5. Contrasting hues efficiently describe qualitative differences on the zoning map shown in monochrome in figure 7.1.

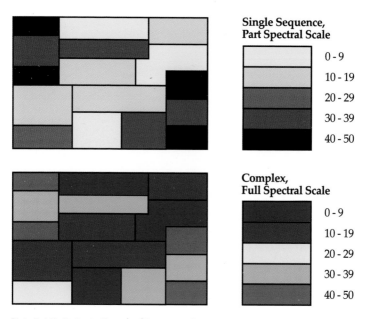

Plate 6. A limited set of hues (top) is more easily grasped than an illogical, complex sequence of spectral hues (bottom).

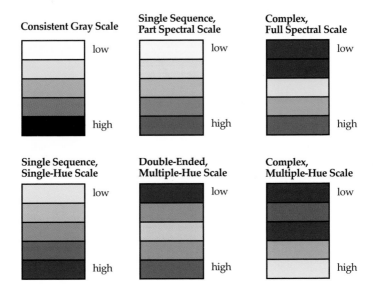

Consistent Gray Scale

low

high

Single Sequence, Part Spectral Scale

low

high

Complex, Full Spectral Scale

low

high

Single Sequence, Single-Hue Scale

low

high

Double-Ended, Multiple-Hue Scale

low

high

Complex, Multiple-Hue Scale

low

high

Plate 7. Some color sequences found on choropleth maps.

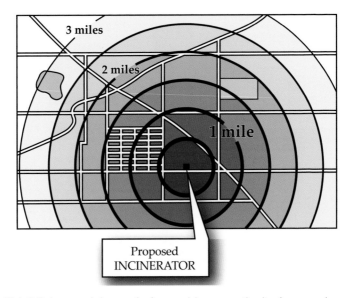

3 miles

2 miles

1 mile

Proposed
INCINERATOR

Plate 8. Red area symbols connoting increased danger near the site of a proposed incinerator strengthen the message of a monochrome environmental-propaganda map (fig. 8.16).

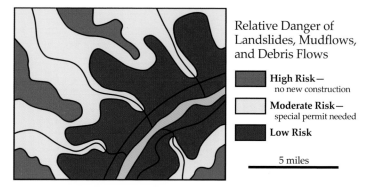

Relative Danger of Landslides, Mudflows, and Debris Flows

High Risk—
no new construction

Moderate Risk—
special permit needed

Low Risk

5 miles

Plate 9. The map viewer reminded of the graphic metaphor can readily decode a sequence of three traffic-light colors portraying degree of environmental risk.

Deasy Hills

x Mt. Raisz

Harley River

Woodward Uplands

Snyder Is.

Relief

Feet above Sea Level

5000
2000
1000
500
0

100 miles

Plate 10. Although hypsometric tints are widely used to portray relief with color-coded elevation categories, map viewers must be aware that lowland areas shown in green might well be dry and barren.

Plate 11. The squares at the centers of these three boxes are identical, but because of simultaneous contrast the gray center appears darker when surrounded by a more brilliant color.

Plate 12. Excerpts from a topographic map (left) and the corresponding orthophoto map (right) for Bath, Maine, underscore the differences between line and image maps.

Plate 13. A color infrared-satellite image showing the path of a tornado. White frames relate an area of detail (lower right) to its position on the image. The black area is outside the satellite's ground swath.

Plate 14. One of several annotated image maps Colin Powell presented to the UN Security Council on February 5, 2003.

Plate 15. A reinterpretation of the slide in plate 14 underscores the ease with which questionable labels can be affixed to image maps.

Plate 16. In late 2012 the White Plains, New York, publication *Journal News* posted a zoomable map on which interactive "roll over" symbols (red dots) identified people with pistol permits.

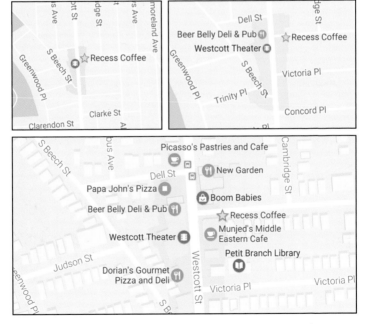

Plate 17. Successively zoomed-in views (upper left → upper right → bottom) of an online street map.

Figure 8.1. Engravings reflect the iconic significance of maps and atlases as national symbols in Christopher Saxton's 1579 *Atlas of England and Wales* (left) and Maurice Bouguereau's 1594 *Le théâtre françoys* (right).

If this map-as-symbol-of-the-state concept seems far-fetched, consider the national atlases that England and France produced in the late sixteenth century. Elizabeth I of England commissioned Christopher Saxton to carry out a countrywide topographic survey of England and Wales and to publish the maps in an elaborate, hand-colored atlas. In addition to providing information useful for governing her kingdom, the atlas bound together maps of the various English counties and asserted their unity under Elizabeth's rule. Rich in symbolism, the atlas's frontispiece (fig. 8.1, left) was a heavily decorated engraving that identified the queen as a patron of geography and astronomy. A few decades later, Henry IV of France celebrated the recent reunification of his kingdom by commissioning bookseller Maurice Bouguereau to prepare a similarly detailed and decorated atlas. Like Saxton's atlas, *Le théâtre françoys* includes an impressive engraving (fig. 8.1, right) proclaiming the glory of king and kingdom. In both atlases regional maps provided geographic detail and a single overview map of the entire country asserted national unity.

The spate of newly independent states formed after World War II revived the national atlas as a symbol of nationhood. Although a few countries in western Europe and North America had state-sponsored national atlases in the late nineteenth and early twentieth centuries, these served largely as reference works and symbols of scientific achievement. But between 1940 and 1980 the number of national atlases increased from fewer than twenty to more than eighty, as former colonies turned to cartography as a tool of both economic development and political identity. In the service of the state, maps and atlases often play dual roles.

Perhaps the haste of new nations to assert their independence cartographically reflects the colonial powers' use of the map as an intellectual tool for legitimizing territorial conquest, economic exploitation, and cultural imperialism. Maps made it easy for European states to carve up Africa and other "heathen" lands, lay claim to land and resources, and ignore existing social and political structures. Knowledge is power, and crude explorers' maps made possible treaties between nations with conflicting claims. That maps drawn up by diplomats and generals became a political reality lends an unintended irony to the aphorism that the pen is mightier than the sword.

Nowhere is the map more a national symbol and an intellectual weapon than in disputes over territory. When nation A and nation B both claim territory C, they usually are at war cartographically as well. Nation A, which defeated nation B several decades ago and now holds territory C, has incorporated C into A on its maps. If A's maps identify C at all, they tend to mention it only when they label other provinces or subregions. If nation B was badly beaten, its maps might show C as a disputed territory. Unlike A's maps, B's maps always name C. If B feels better prepared for battle or believes internal turmoil has weakened A, B's maps might more boldly deny political reality by graphically annexing C.

Figure 8.2. Disputed India-Pakistan boundary and the territory of Jammu and Kashmir, as portrayed in the 1965 *Area Handbook for Pakistan*, published by the US government.

Neutral countries tread a thin cartographic line by coloring or shading the disputed area to reflect A's occupation and perhaps including in smaller type a note recognizing B's claim. If A and B have different names for C, A's name appears, sometimes with B's name in parentheses. (Even when recapture by B is improbable, mapmakers like to hedge their bets.) Cartographic neutrality can be difficult, though, for customs officials of nation B sometimes embargo publications that accept as unquestioned A's sovereignty over C. If A's rule is secure, its censors can be more tolerant.

Consider, for example, the disputed state of Jammu and Kashmir, lying between India, Pakistan, and China. Both India and Pakistan claimed Kashmir, once a separate monarchy, and went to war over the area in August 1965. Figure 8.2, a US State Department map, shows the cease-fire line of fall 1965, which placed Pakistan in control of northwestern Kashmir and showed India in control of the southern portion. (China occupied a portion of northeastern Kashmir.) Nonetheless, Indian and Pakistani government tourist

Figure 8.3. Subtle and not-so-subtle cartographic propaganda on Argentinian postage stamps.

maps have continued to deny political reality, with Pakistani maps including Kashmir in Pakistan and their Indian counterparts ceding the entire territory to India. American and British atlases have attempted to resolve the dispute with notes identifying the area occupied by Pakistan and claimed by India, the area occupied by India and claimed by Pakistan, three areas occupied by China and claimed by both India and Pakistan, the area occupied by China and claimed by India, and the area occupied by India and claimed by China. And for years publishers have found it difficult to export the same books on South Asian geography to both India and Pakistan.

Even tiny maps on postage stamps can broadcast political propaganda. Useful both on domestic mail to keep aspirations alive and on international mail to suggest national unity and determination, postage-stamp maps afford a small but numerous means for asserting territorial claims. As shown in figure 8.3, Argentinian postage stamps have touted that nation's claims not only to the Falkland Islands and the British-held islands to their east but also to a slice of Antarctica. Like all official maps of Argentina, these postage stamps deny the legitimacy of British occupation with their Spanish label for the Falkland Islands: "Islas Malvinas." Postage stamps bearing maps are also useful propaganda tools for emergent nations and ambitious revolutionary movements.

Size, Sympathy, Threats, and Importance

Sometimes propaganda maps try to make a country or region look big and important, and sometimes they try to make it look small and threatened. In the former case, the map might support an appeal to fairness: the developing area is big, and therefore it deserves to consume a larger share of the world's resources, exercise more control over international political bodies such as UNESCO (the United Nations Educational, Scientific, and Cultural Organization), and receive greater respect and larger development grants from the more developed nations of the world. In the latter case, the map might dramatize the threat a large state or group of states poses for a smaller country. Figure 8.4, for instance, portrays a cartographic David-and-Goliath contest between tiny Israel and the massive territory of the nearby oil-rich Arab nations. Even though the map's geographic facts are accurate, a map comparing land area tells us nothing about Israel's advanced technology, keen military preparedness, and alliances with the United States and other Western powers.

Some map projections can help the propagandist by making small areas bigger and large areas bigger still. No projection has been as abused in the pursuit of size distortion as that devised by sixteenth-century atlas publisher and cartographer Gerardus Mercator. Designed specifically to aid navigators, the Mercator projection vastly enlarges poleward areas so that straight lines can serve as *loxodromes*, or *rhumb lines*—that is, lines of constant geographic direction. (If the navigator's compass shows true north rather than magnetic north, rhumb lines can be called lines of constant com-

Figure 8.4. Map showing the encirclement of Israel by neighboring Arab nations, redrawn from a map published during the 1973 war by the Jewish National Fund of Canada.

pass direction.) As figure 8.5 shows, the navigator finds the course by drawing a straight line from origin A to destination B and then reading the angle θ from the meridian to the rhumb line. If one consistently follows this bearing from A, one will eventually reach B. For this convenience the navigator must sacrifice a shorter but less easily followed great-circle route and endure the areal distortion caused by the progressive increase poleward of north-south scale. In fact, the projection shows little of the area within the Arctic Circle and the Antarctic Circle because its poles are infinitely far from its equator. Ever wary of icebergs anyway, navigators for centuries have avoided polar waters and accepted as only a minor liability the Mercator projection's gross areal exaggeration. Yet for decades the John Birch Society and other political groups warning of the "red menace" commonly shared the stage with a massive Mercator map of the world with cartographically enhanced China and Russia printed in a provocative, symbolically rich red.

Although equal-area map projections (as in figs. 2.5 and 2.6) have been available since at least 1772, when Johann Heinrich Lambert published his classic *Beiträge zum Gebrauche der Mathe-*

Figure 8.5. Mercator world map showing the bearing angle θ for a rhumb line from A to B and the areal exaggeration of Red China and in particular the USSR. Designed to aid navigators, the Mercator also has served political propagandists seeking to magnify the Communist threat.

matik und deren Anwendung, Mercator's projection provided the geographic framework for wall maps of the world in many nineteenth- and early twentieth-century classrooms, and later for the sets of television news programs and backdrops of official briefing rooms. (It's also well suited for and widely used in interactive online maps; see chapter 14.) Perhaps distracted by concerns with navigation, exploration, and time zones, cartographically myopic educators and set designers presented a distorted worldview that diminished the significance of tropical areas to the advantage of not only Canada and Siberia but western Europe and the United States as well. The English especially liked the way the Mercator flattered the British Empire with a central meridian through Greenwich and prominent far-flung colonies in Australia, Canada, and South Africa. Some British maps even gave the empire an added plug by repeating Australia and New Zealand at both the left and right sides of the map.

Yet in the early 1970s this subtle and probably unwitting geopo-

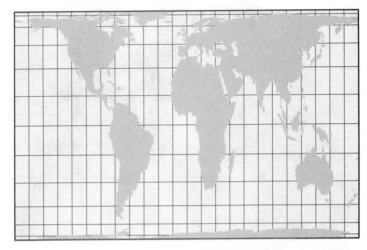

Figure 8.6. The Peters projection or, more accurately, the Gall-Peters projection.

litical propaganda served as a convenient straw man for German historian Arno Peters, who published a "new" world map based on an equal-area projection similar to one described in 1855 by the Reverend James Gall, a Scottish clergyman. As figure 8.6 shows, the Gall-Peters projection gives tropical continents a mildly attenuated, stretched look, which probably explains why geographers and cartographers have adopted more plausible equal-area maps and why the basic texts on map projections that Peters had consulted ignored Gall's contribution. Indeed, Lambert and other cartographers had developed numerous equal-area map projections, including many that distort shape much less severely than the Gall-Peters version does.

But Dr. Peters knew how to work the crowd. A journalist-historian with a doctoral dissertation on political propaganda, Peters held a press conference to condemn the Mercator worldview (as well as all nonrectangular projections) and to tout his own projection's "fidelity of area" and more accurate, "more egalitarian" representation of the globe. By calling attention to the Mercator's slighted portrayal of most developing nations and blaming a stagnation in the development of cartography, Peters struck

Figure 8.7. Like all equal-area projections, this hourglass equal-area map projection John Snyder devised as a joke has area fidelity but distorts shape.

responsive chords in the World Council of Churches, the Lutheran Church of America, and various United Nations bodies. Religious and international development organizations welcomed Peters and his "new cartography," with the greater fairness and accuracy it promised. They also published large and small versions of the Peters projection, hung it on their walls, and used it in their press releases and publications. Perhaps because journalists also like to champion the oppressed and can't resist a good fight, the press repeated Peters's claims and reported the success of his bandwagon. Academic cartographers became both puzzled and enraged— puzzled that the media and such prominent, respected institutions could be so gullible and ignorant, and enraged that these groups not only so persistently repeated Peters's preposterous assertions but so obstinately refused to look at cartography's writings, accomplishments, and rich history.

Not all cartographers lacked a sense of humor. US Geological Survey cartographic expert John Snyder, himself a developer of several useful as well as innovative map projections, offered yet another equal-area projection to underscore his cartographic colleagues' point that an equal-area map is not necessarily a good map. As shown in figure 8.7, Snyder's hourglass equal-area projection does what the Peters projection does and the Mercator doesn't—it preserves areal relationships. But it also demonstrates dramatically that areal fidelity does not mean shape fidelity.

Ironically, by succumbing to Peters's hype, UNESCO and other organizations sensitive to problems of the developing world loyally backed the wrong projection and missed an enormous propaganda opportunity. By accepting uncritically the rather dubious assumption that a map responsive to people should accurately represent land area, these groups not only demonstrated a profound cartographic naïveté but also ignored a more humanistic type of map projection that actually makes some developing-world populations appear justifiably enormous, rather than gaunt and emaciated. How much more convincing their media blitz might have been had they supported a demographic base map, or area cartogram, similar to figure 2.10, on which the area of each country is scaled according to number of inhabitants. Indeed, an area cartogram would be more effective than the Peters projection in boosting the importance of China, India, and Indonesia and in revealing the less substantial populations of Canada, the United States, Russia, and other comparatively less crowded countries. But perhaps a subtler internal need motivated leaders of UNESCO and the World Council of Churches, for the Peters projection is comparatively kinder to the low and moderate population densities of Africa, Latin America, and the Middle East—indeed, a cynic might note the influence of African diplomats in UNESCO and the inherent interest of the World Council of Churches in concentrated Christian missionary activity in Latin America and central Africa.

Propaganda Maps and History: In Search of Explanation and Justification

Although propaganda cartography is probably not much younger than the map itself, the Nazi ideologues who ruled Germany from 1933 to 1945 warrant special mention. No other group has exploited the map as an intellectual weapon so blatantly, so intensely, so persistently, and with such variety. Nazi propaganda

Figure 8.8. "Then and Now! 1914 and 1939" (*Facts in Review* 1, no. 17 [December 8, 1939]: 1).

addressed especially to the United States presented a selective and distorted version of history designed to increase sympathy for Germany, decrease support for Britain and France, and keep America out of World War II, at least until Axis forces had conquered Europe. The examples discussed in this section are from the weekly news magazine *Facts in Review*, published in New York City during 1939, 1940, and 1941 by the German Library of Information.

1914—1918: German Fleet Bottled Up.   1940: Germany Breaks Through to the Atlantic.

Figure 8.9. "The War in Maps" (*Facts in Review* 3, no. 16 [May 5, 1941]: 250).

Nazi cartographic propaganda often introduced the sympathy theme by recalling Germany's defeat in World War I—a humiliation followed by an economic depression that helped the National Socialists to power. Figure 8.8, which compares the German plight in 1914 with that of 1939, invokes a persistent anti-British theme. These two maps form much of the front page of *Facts in Review* for December 8, 1939. In the magazine, a caption to the left of the 1914 map notes the encirclement that "provided the necessary basis for Britain's successful Hunger-blockade," whereas the caption for the 1939 map alludes to Britain's failed attempts to repeat the encirclement and proclaims that "the path of industrial and economic cooperation to the East and the Southeast lies open!" Note, though, that the 1939 map conveniently groups Germany's main allies at the time, Mussolini's Italy and Stalin's Russia, with Switzerland and other "neutral countries."

In early 1941, another map attempted to explain and justify Germany's western advance against England into France, Belgium, and Holland by comparing Germany's strategic disadvantage in

A STUDY IN EMPIRES

GERMANY

GREAT BRITAIN

THE AGGRESSOR NATION?

26% OF THE WORLD
(THE BRITISH EMPIRE)

Figure 8.10. "A Study in Empires" (*Facts in Review* 2, no. 5 [February 5, 1940]: 33).

1914 with the more favorable situation in 1940. Figure 8.9 contrasts the German navy "bottled up" by the British in the North Sea in 1914 through 1918 with the German navy that in 1940 had "[broken] through to the Atlantic." Hitler had not yet turned against Stalin, and in the magazine the map's caption notes that whereas Germany had to fight on two, and later three, fronts in 1914, "Today no such danger exists. The British blockage is ineffective and, instead, the blockaders themselves are being blockaded." Arcs reinforce the blockade theme of the 1914–1918 map, and bold arrows dramatize Germany's freer access to the Atlantic on the 1940 map.

Other Nazi maps attempted to divert sympathy from Britain. Captioned "A Study in Empires," the charts in figure 8.10 compare the 264,300 square miles on which Germany's 87 million inhabitants "must subsist" with the 13,320,854 square miles that Britain, with only 46 million people, "has acquired." How can little Germany be the aggressor nation? the left panel asks. In contrast, the right

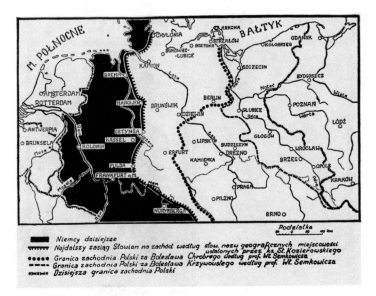

Figure 8.11. "Polish Delusions of Grandeur" (*Facts in Review* 2, no. 28
[July 8, 1940]: 294).

panel suggests a note of greed in Britain's conquest of 26 percent of
the world's land area. The map's caption sounds a further chord of
grievance by noting that the British Empire "includ[es] the former
German colonies."

Another plot revealed in *Facts in Review* justified the partition
of Poland between Germany and Russia. Captioned "Polish Delusions of Grandeur," figure 8.11 shows in bold black a much-reduced
German state. Offended and outraged, the editors reveal that "this
map, published in the Posen newspaper, 'Dziennik Poznanski,'
after the receipt of Chamberlain's 'blank check,' revived dreams
of extending the Polish dominion to the Weser River." Although
a newspaper map hardly constitutes official state policy, the map
suggests to the politically naive that the 1939 invasion amply repaid
the Poles for even daring to think of annexing German territory.

Useful for representing one's opponents as the bad guys, maps
can also advertise oneself as the good guy. Accompanying a story
headlined "Repatriation: Background for Peace," figure 8.12 shows

Figure 8.12. "Repatriation: Background for Peace" (*Facts in Review* 1, no. 16 [November 30, 1939]: 3).

Germany the Peacemaker quietly reducing ethnic friction in the Baltic states by evacuating 80,000 to 120,000 Germans. As *Facts in Review* proudly observes, "Germany is not afraid to correct mistakes of geography and history." The map's pictorial symbols dramatize the repatriation by showing proud, brave, obedient Germans clutching their suitcases and lining up to board ships sent to "lead [these] lost Germans back home to the Reich." To the east in stark, depressing black looms the Soviet Union, and to the south in pure, hopeful white lies Germany.

Figure 8.13. "Spheres of Influence" (*Facts in Review* 3, no. 13 [April 10, 1941]: 182).

In trying to persuade the United States to remain neutral, Nazi cartographic propagandists flattered both isolationism and Monroe Doctrine militarism. Titled "Spheres of Influence," figure 8.13 uses bold lines to send a clear message to Americans: stay in your own hemisphere and out of Europe. Faintly resembling the lobes of Goode's interrupted projection (fig. 2.6), familiar to many students, the map also supported a geopolitical theater for Germany's Pacific ally, Japan. How successful the Nazi cartographic offensive might have been is moot, for the United States entered the war on the side of England after Japan attacked Pearl Harbor, Hawaii, on December 7, 1941.

Arrows, Circles, Place-Names, and Other Cartographic Assault Weapons

Few map symbols are as forceful and suggestive as the arrow. A bold, solid line might make the map viewer infer a well-defined, generally accepted border separating neighboring nations with homogeneous populations, but an arrow or a set of arrows can dramatize an attack across the border, exaggerate a concentration

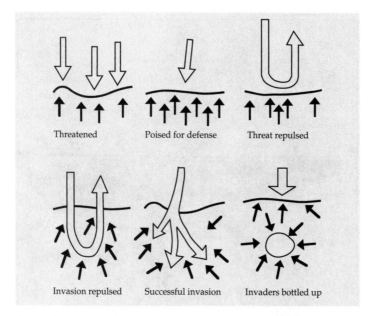

Figure 8.14. Arrow symbols portraying a variety of maneuvers and stalemates.

of troops, and perhaps even justify a "preemptive strike." As figure 8.14 demonstrates, arrow symbols can vary in size, number, and arrangement to portray a range of military confrontations, from overwhelming threats and courageous standoffs to invasions with varying degrees of success. During World War II and the Korean War, many American newspapers used daily battlefield maps with forceful and suggestive arrows to give their readers a generalized blow-by-blow account of the Allied forces' victories and defeats. As figure 8.15 demonstrates, prominent arrows and black areas portraying captured territory could dramatize the threat of an advancing enemy.

A less abstract cousin of the arrow is the bomb or missile symbol. Everybody knows what it is and fears its referent. Lines of miniature missiles and stacks of ominous little red or black bombs readily impress map viewers with the comparative sizes of opposing arsenals. Orientation is also important: bombs are stockpiled

Figure 8.15. A 1950 Associated Press newspaper map uses black shading to mark the part of South Korea that was invaded by North Korean forces and uses arrows to portray troop movements.

horizontally but dropped vertically, whereas missiles are stored upright but hurled horizontally. To justify an expanding defense budget, a propagandist might even stage a mini nuclear attack, complete with a victorious response. Maps can even make nuclear war appear survivable.

The specter of nuclear warfare sends threatened nations and pacifists worldwide to the cartographic arsenal for an honored piece of geopolitical ordnance, the circle. Diplomats and military strategists have found the circle particularly useful in showing the striking zones of aircraft, and modern strategists find circles indispensable when discussing the range of guided missiles. Circles bring to the map a geometric purity easily mistaken for accuracy

Figure 8.16. A local environmental-protection group might seek to arouse citizen support with a propaganda map on which concentric circles have progressively more-threatening labels closer to the site of a proposed incinerator.

and authority. Yet on few small-scale maps do circles on the sphere remain circles in a two-dimensional plane. Even local environmental activists find circles useful, especially when arranged concentrically around the site of a proposed incinerator or nuclear-power plant, and with ever larger, more threatening labels for closer circles, as in figure 8.16 (plate 8).

Naming can be a powerful weapon of the cartographic propagandist. Place-names, or *toponyms*, not only make anonymous locations significant elements of the cultural landscape but also offer strong suggestions about a region's character and ethnic allegiance. Although many maps not intending a hint of propaganda might insult or befuddle local inhabitants by translating a toponym from one language to another (as from Trois Rivières to Three Rivers) or by attempting a phonetic transliteration from one language to another (as from Moskva to Moscow) or even from one alphabet to another (as in Peking or Beijing), skillful propagandists have often altered map viewers' impressions of multiethnic cultural land-

Figure 8.17. Dramatic map comparing infant-mortality rates for parts of San Diego, California, with national rates of various countries.

scapes by suppressing the toponymic influence of one group and inflating that of another.

Local social activists can also use the suggestive power of place-names to make a point cartographically. Figure 8.17, for instance, is an infant-mortality map of San Diego, California, that strongly indicts intra-urban inequalities in maternal and infant health care. As the map notes, some parts of the city are comparable to highly developed western European nations such as Sweden and Switzerland, whereas other neighborhoods are similar to Hungary or Jamaica. Figures 8.16 and 8.17 both demonstrate that cartographic propaganda can be an effective intellectual weapon against an unresponsive, biased, or corrupt local bureaucracy.

Like guns and lacrosse sticks, maps can be good or bad, depending on who's holding them, who they're aimed at, how they're used, and why.

CHAPTER 9

Maps, Defense, and Disinformation: Fool Thine Enemy

Compared with military maps, most propaganda maps are little more than cartoons. A good defense establishment knows how to guard its maps and their geographic details and yet at times leak false information that the enemy might think is true. Providing some accurate information is necessary, of course, if the "disinformation" is to be credible. An intellectual weapon in political propaganda, the map has been a fundamental tactical weapon for military counterintelligence and covert diplomacy.

This chapter addresses how and why governments guard maps, hide geographic information, and sometimes even distribute deliberately falsified maps. The first section discusses the very real need for cartographic security, the second examines the now-admitted excesses of cartographers in the former Soviet Union who deliberately doctored their maps, and the third section explores how governments sometimes mislead their own citizens by failing to include threats to a sound environment and other possible embarrassments in their maps.

Defense and a Secure Cartographic Database

No doubt about it: mapped information often must be guarded. If knowledge is power, an enemy's knowledge of your weaknesses and strengths is a threat. Maps can also betray your plans, as Giovanni Vigliotto discovered. In 1981 an Arizona jury found this fifty-three-year-old ladies' man guilty of fraud and bigamy. Giovanni, who claimed to have married more than 105 women over thirty-three years, invariably cut short each honeymoon by absconding with his victim's cash and jewelry. Had he not left behind an annotated map when he abandoned one of his wives, Giovanni might not have been caught. Vehicle navigation systems can leave a similar trail.

Nations too try to keep maps out of enemy hands, even obsolete maps. In 1668, Louis XIV of France commissioned three-dimensional scale models of eastern border towns, so that his generals in Paris and Versailles could plan realistic maneuvers. On exhibit in Paris in La Musée des Plans-Reliefs, at the Hôtel des Invalides, these highly detailed wood and silk models are amazingly accurate portraits of seventeenth-century French towns. As late as World War II, the French government guarded them as military secrets with the highest security classification, but nowadays they're an intriguing tourist attraction. Less cartographically ambitious states also tightly control battle plans and maps that contain strategic information about communications, fortifications, and transport. Giving the enemy a detailed map has often been considered an act of treason, unless of course the map itself is a fraud designed to confuse the opponent or persuade him to attack or not attack.

What is mapped as well as the maps themselves must be kept confidential, for to reveal an interest in a particular area or feature is to reveal one's plans. Governments compile maps of foreign areas in secure, windowless buildings using cartographers with "secret" or "top secret" security clearances because they don't want enemies, neutral countries, or even allies to know what interests them.

Mapping agencies must, of course, guard against fire, natural disasters, and sabotage, but the electronic age has brought a new set of concerns. In the 1970s electronic maps stored in computer databases began to replace traditional maps on paper or plastic film as the principal format for organizing and storing geographic data, including sensitive military information. Such maps are vulnerable to computer hackers and to a type of nuclear attack called an *electromagnetic pulse*. Hackers sometimes alter information or develop computer viruses that destroy data and programs, possibilities that were unknown before telecommunication net-

works allowed computers to talk to each other as well as to distant
users. Known to have penetrated supposedly secure Department of
Defense computers, hackers seem intrigued by the challenge and
thrill of entering a cartographic data system and moving Montana or flooding the Dead Sea. More threatening than the malicious lone hacker is the military or terrorist organization poised
to launch a cyberattack.

A second threat to maps stored on magnetic media, the electromagnetic pulse (EMP) is a burst of radiation hurled downward by
a high-altitude thermonuclear explosion or a massive solar flare.
Able to destroy power and telecommunications transmission systems and damage integrated circuits, optical fibers, and magnetic
storage devices, this sudden, intense radiation could render unreadable most maps published online or viewed on electronic devices. To
protect themselves against an EMP, governments have attempted to
"harden" their electronic information systems and stockpile maps
on paper, microfilm, and other nonmagnetic media. Should civilization endure and survive a nuclear attack, the traditional, bulky
cartographic image would be the ultimate backup for the more flexible yet more vulnerable electronic map.

Soviet Cartography, the Cold War, and Displaced Places

Keeping all or even most maps away from enemy eyes is a nearly
impossible task. A still greater hindrance to map users, though,
is deliberate, widespread cartographic "disinformation." While
difficult to pull off today, this was a favored Cold War tactic of the
former Soviet Union.

Although other nations might have also intentionally distorted
their maps, the USSR's systematic falsification of geographic locations is now a well-known part of the recent history of cartography.
In the late 1930s, after the NKVD, or security police, assumed control of mapmaking, the Soviet cartographic bureaucracy began to
deliberately distort the positions and forms of villages, coastlines,
rivers, highways, railroads, buildings, boundaries, and other fea-

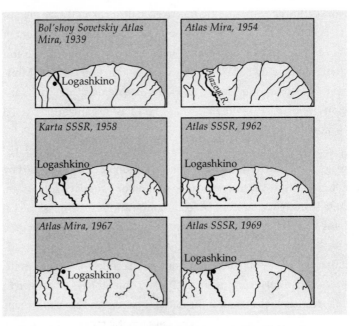

Figure 9.1. Representation of Logashkino and vicinity, on the East Siberian Sea, on various Soviet maps published between 1939 and 1969.

tures shown on maps and atlases sold for public use. This policy reflects a police-state mentality not unlike the misguided disinformation campaigns and cover-ups that at times have embarrassed the United States, Britain, and other Western governments. Apparently the Soviets wanted military planners in China and the West to consider their maps untrustworthy so they would not record coordinates that might be used for guiding cruise missiles to Soviet sites.

How severe were the distortions? Figure 9.1 provides an example in the puzzling peregrinations of the town of Logashkino, on or near the shore of the East Siberian Sea, as portrayed on various Soviet maps published between 1939 and 1969. The *Bol'shoy Sovetskiy Atlas Mira*, published in 1939, showed Logashkino on the Alazeya River, well inland from the coast, but the *Atlas Mira* published in 1954 omitted the town altogether and showed the river with only a single channel. *Karta SSSR* for 1958 recreated Logashkino on the coast

Figure 9.2. Representation of Salmi and vicinity, on Lake Ladoga near 32° E, on Soviet maps published between 1962 and 1969.

and restored the river's other channel. The *Atlas SSSR* published in 1962 offered a similar portrayal of the town and the river, but *Atlas Mira* for 1967 eliminated the eastern channel and moved Logashkino inland. Finally, the 1969 edition of *Atlas SSSR* again moved the town to the coast and showed both channels. The anonymous writer who described these mercurial representations for the magazine *Military Engineer* wryly observed, "Apparently there is such a town, but whether it is on the seacoast or on a river, or neither, is a matter of uncertainty when based on the work of the Soviet cartographers."

Displacement was particularly noticeable for towns near a meridian or parallel. Figure 9.2 illustrates the coordinate shifts for Salmi, a town on the north shore of Lake Ladoga, and a nearby offshore island. The heavy vertical line through each panel is the meridian 32° E. Although the *Atlas SSSR* shows Salmi about 10 kilometers west of the meridian in 1962, the *Atlas Mira* edition of 1967 moves the town eastward, directly astride the meridian. And *Atlas SSSR* for 1969 shifts the town 4 kilometers farther east. The 32nd meridian, which on the 1962 map falls well east of the island, nearly cuts the island in half on the 1969 map. Although Salmi moved only 14 kilometers, or nearly 9 miles, between 1962 and 1969, Soviet maps have displaced other towns by as much as 25 miles.

Soviet cartographic disinformation even affected tourist maps of urban areas. Detailed street maps of Moscow and other Soviet cities often failed to identify principal thoroughfares and usually omitted a scale, so that distances were difficult to estimate.

Figure 9.3. Portion of Griffiss Air Force Base, in Rome, New York, as shown on a large-scale planimetric map sold to the public by the state of New York.

Although local citizens were well aware of its presence, Soviet street maps of Moscow suppressed the imposing KGB building on Dzerzhinski Square, as well as other important buildings. In contrast, the CIA pocket atlas used by American foreign-service personnel had a detailed, fully indexed, easy-to-use map that showed the KGB headquarters and other important landmarks. Imagine the outcry if Soviet diplomats had had better maps of Washington, DC, than US government workers, tourists, and other citizens.

So why did Soviet cartographers stop fudging their maps?

First, cartographic disinformation is costly both to mapmaking and to economic development. To be convincing, feature falsification takes time and personnel better used to make the nation's maps more accurate and up-to-date. Maintaining a second, secure set of accurate maps is expensive, and providing carefully controlled access to economic planners and other decision makers is costly, time-consuming, and risky. Second, spy satellites have made fudged maps less useful and surely less necessary than in the 1960s and earlier, when the USSR's enemies still depended largely on existing old maps, spies, defectors, and U-2 spy planes. Modern intelligence satellites provide routine land-cover surveillance of potentially hostile countries, and image-analysis computer systems scan satellite imagery in a vigilant search for suspicious changes in land cover. Some intelligence satellites with high-resolution sensors can even alter their orbits to collect highly detailed imagery of suspect areas, and resolution adequate for discerning the make and model of a motor vehicle is more than adequate for monitoring missile launch sites and troop movements. Indeed, the USSR's cartographic obfuscations seem quaintly absurd a half century later, when anyone in most parts of the world can go online and zoom in on every Russian city.

Features Not Shown, Maps Not Made

Are other countries' maps of sensitive sites more open, more revealing, more complete than Cold War–era Soviet maps? Generally, yes. For example, figure 9.3, a large-scale state-highway-department map sold to the public in the 1970s, shows considerable ground detail for Griffiss Air Force Base in the city of Rome, New York. Home to long-range bombers of the Strategic Air Command (SAC) and advanced research projects on electronic aerial reconnaissance at the Rome Air Development Command (RADC), Griffiss would have been a prime target in a war with the Soviet Union.

Yet cartographic openness was not universal in the United States, although the exceptions are often more puzzling than frus-

The presidential retreat, **Camp David**, is in the middle of the park. It is closed to the public, but its entrance and some of the maximum security fencing can be seen to the right on Park Central Road, north of the visitor center on Md. 77.

Figure 9.4. Camp David, a portion of Catoctin Mountain National Park, in Frederick County, Maryland, appears anonymously as "Camp 3" on a US Geological Survey topographic map of the area (above). In contrast, a guidebook provides concise directions (below).

trating. As an illustration, figure 9.4 shows how maps of Catoctin Mountain National Park in western Maryland weakly camouflaged Camp David, the famous presidential retreat, with the nondescript label "Camp 3." Yet, surprisingly, the map also offers a few clues, such as the perimeter road along the security fence. As quoted at the bottom of the figure, a 1976 guidebook for highway travelers in Maryland readily thwarted this feeble attempt at geographic anonymity by providing accurate directions to the camp. More recently, online mapping applications and freely available overhead imagery have further blown the camp's cover by revealing diverse improvements to the site's recreational and helicopter facilities.

This widespread openness was challenged in late 2001 after terrorists attacked the World Trade Center and the Pentagon with hijacked aircraft. Public officials in charge of data-sharing centers called for drastic restrictions on access to detailed geographic data, which could, they argued, help terrorists select targets and plan attacks. Details of industrial facilities were blurred on digital aerial imagery, and map libraries were asked to deny access to public data to anyone not a serious, credentialed researcher. This hysteria faded after a study funded by the Department of Defense concluded that sensitive information was readily available through other channels and that restricting access to publicly funded geographic information would adversely affect science, education, economic development, and public administration.

Historically, Britain seems to have been more cartographically paranoid than the United States, according to the *New Statesman*, a London journal of left-leaning political commentary. Britain's national civilian mapping agency, the Ordnance Survey, apparently maintained lists of sensitive sites that had to be omitted or disguised on maps and somehow camouflaged on aerial photographs during the Cold War. Thus nuclear bunkers masqueraded as warehouses, while radio stations, plants processing nuclear fuel, and government-owned oil depots vanished altogether. Some countries, such as Greece, have published maps with large, telltale blank areas—less devious, perhaps.

Americans need not applaud themselves for their comparative openness. After all, British official maps tend to show more detail at larger scales than their American counterparts. Moreover, American maps often omit information that might embarrass industrial polluters or local officials. For example, figure 9.5 shows two maps centered on the infamous Love Canal, a neighborhood near Niagara Falls, New York, that is contaminated by hazardous waste. The 1946 map, which shows the canal as a long, straight vertical feature, fails to indicate use of the canal since 1942 as a dump for chemical waste. The 1980 map not only shows no trace of the filled-in canal but ignores the area's tragic history: dumping continued until 1953;

Figure 9.5. Love Canal area in Niagara Falls, New York, as shown on large-scale topographic maps from 1946 (left) and 1980 (right).

developers filled in the canal and built homes there in the 1950s; the city built a public elementary school across the filled-in canal in 1954; and chemical seepage spread up to the surface and laterally into the basements of nearby homes. After people and pets became ill in the early 1970s, soil analyses revealed abnormally high concentrations of chlorobenzene, dichlorobenzene, and toluene, and in 1978 the New York health commissioner declared a state of emergency and relocated 239 families. Although federal and state mapping agencies contended that topographic maps should show only standardized sets of readily visible, more-or-less permanent features, such assertions seem hypocritical when these agencies' maps routinely included boundary lines, drive-in movie theaters, and other elements far less important to human health.

As historian of cartography J. B. Harley has noted, government maps have for centuries been ideological statements rather than fully objective, "value-free" scientific representations of geographic reality. Harley observed that governments practice two forms of cartographic censorship—a censorship of secrecy to serve military defense and a censorship of silence to enforce or reinforce social and political values. Social-political censorship can assert the power of the state or the rights of private landowners, and it can also attempt to calm ethnic minorities, as in 1988 when New York governor Mario Cuomo decreed that all derogatory place-names be stricken from state maps. But as with the Love Canal maps, this second, more subtle form of cartographic censorship usually occurs as silences—as ignored features or conditions. Hence basic maps of most cities show streets, landmark structures, elevations, parks, churches, and large museums—but not dangerous intersections, impoverished neighborhoods, high-crime areas, and other zones of danger and misery that could be accommodated without sacrificing information about infrastructure and terrain.

By omitting politically threatening or aesthetically unattractive aspects of geographic reality and by focusing on the interests of civil engineers, geologists, public administrators, and land developers, our topographic "base maps" are hardly basic to the concerns of public health and safety officials, social workers, and citizens rightfully worried about the well-being of themselves and others. In this sense, cartographic silences are indeed a form of geographic disinformation.

Large-Scale Mapping, Culture, and the National Interest

Lest the reader be left with the impression that agencies and companies responsible for large-scale maps are either deliberately devious or flagrantly irresponsible, this chapter explores the influence of culture on topographic mapping. Always a loaded concept for social scientists, culture accounts for many routine, seemingly automatic decisions about features and their portrayal as well as for national policy on collecting and disseminating cartographic information. As we'll see, our nation's topographic maps reflect widely shared Western values (geometric precision, consistency, completeness, cost-effectiveness) as well as the professional subcultures of mapmaker-bureaucrats and, more recently, online map providers. A few international comparisons demonstrate that the American way, however workable, is not the only way.

Focused primarily on the operations and mores of national mapping organizations, in particular the United States Geological Survey (USGS), this chapter offers insights to occasional users of topographic maps as well as to businesses and local governments that are heavily dependent on cartographic information. After conducting a cursory examination of the national mapping enterprise and its evolution, the first section looks at how rigorously defined feature types, map symbols, and generalization procedures promote consistency among the hundreds or thousands of individual quadrangle maps that constitute a map series. The second section, on cartographic agendas, explores mapmaking as an institutional process subject to bureaucratic norms, administrative red tape, and political pressure. A short concluding section notes the emergence of private-sector online mapping, which preempts or extends several roles of the national topographic map.

Standards and Specifications

Nothing better reflects the government cartographer's bureaucratic mentality than the standards and specifications for a nationwide series of topographic maps. In partitioning an entire country among a largely arbitrary grid of rectangular areas called *quadrangles*, the national mapping organization willingly sacrifices political, ethnic, and physical boundaries to the convenience of uniformly spaced meridians and parallels—a divide-and-conquer strategy that makes complete coverage seem both doable and essential. Even though the mapmaker is not a foreign invader, the interests of local communities divided among two or more quadrangles crumble in the path of a cartographic enterprise committed to scientifically higher, geographically broader needs.

What are these needs? National mapping as practiced in the United States and other developed countries has long served complementary goals of national defense and economic development. Military leaders need detailed maps to identify points of attack and plan their defense. Since an effective defense requires not only planning but rapid deployment of troops and weapons, a nation also needs roads, railways, and dependable waterways—infrastructure that is equally useful in getting raw materials to factories and manufactured goods to market. Topographic surveys, which help civil engineers lay out routes and build bridges, provide the geometric foundation for maps on which transportation structures are basic features. Maps, of course, also promote mineral exploration and agricultural development, and support private ownership—and taxation—of land. And as a consequence of economic development, industrialized societies need maps for environmental protection and growth management. However specific a nation's mapping requirements, most series of large-scale maps address several, if not all, of these four principal objectives.

A geographic windfall of sorts, large-scale maps afford a base for compiling a wide variety of smaller-scale products, including road maps, real-estate maps, and maps in books, scientific papers, and news media. In 1892, Henry Gannett, the Geological Survey's first chief topographer, recognized the importance of detailed carto-graphic sources by calling his department's 1:125,000 and 1:62,500 quadrangle maps "mother maps." An admirer of the systematic large-scale map series produced by national surveys in Europe, Gannett sought a comparable treatment of the United States.

In selecting cartographic features, early USGS mapmakers gave highest priority to "natural features" such as streams and land-forms. Among the "artificial features" of the built environment, they distinguished "public culture," such as canals, railroads, and wagon roads, from "private properties," such as houses, fields, and orchards. Public topography was worth showing, but private topog-raphy, deemed less permanent, could require constant revision at great cost as well as interfere graphically with symbols represent-ing natural and public features. Even so, some private topography (dwellings, larger commercial buildings) seemed an essential part of the country's basic geographic frame of reference. And in the 1940s, a new series of 1:24,000 maps, based on 7.5-minute quadrangles each covering only a fourth as much territory as existing 1:62,000, 15-minute maps, allowed more room for showing homes, factories, sewage treatment plants, power lines, and other features collective-ly identified in cartographic jargon as "culture."

In addition to specifying which natural and built features to include, USGS officials prepared a manual to guide workers throughout the agency in both generalization and symbolic repre-sentation. Take, for example, the water gap, a deep, narrow break cut by a river through an otherwise continuous linear ridge. People often build roads and railways through water gaps. As illustrated in figure 10.1, crowded features challenge mapmakers to maintain clarity and legibility while packing a narrow graphic corridor with symbols portraying roads, railways, riverbanks, and their interior flow lines as well as contour lines describing slope and elevation.

Figure 10.1. Because visibility usually requires road and railway symbols to be wider than the map scale calls for, drainage symbols and contour lines have been displaced laterally to minimize graphic congestion on this 1902 1:62,500 USGS topographic map (enlarged to about 200 percent of original scale) showing a water gap just west of Everett, Pennsylvania.

To promote consistency within a map series and prevent disagreement between workers and supervisors, national mapping organizations established rules for dealing with cartographically contentious situations. In the case of water gaps, a prescribed order of inking—culture first, then drainage, and then topography—not only avoided graphic congestion by displacing contours and streams but obviated awkward kinks in track and highway alignments.

Standards and specifications have become ever more exacting, with rules devised by committees telling mapmakers how to select features and show details. As the following USGS road-selection rules illustrate, size is significant, but relevance to humans can make small features important:

> Private roads, access roads, and driveways less than 500 feet (152.4 m) in length will not be shown unless of landmark value in areas of sparse culture.

All streets in populated places will be shown regardless of length.

And as these guidelines for portraying railways point out, a feature's overall form can be more relevant than its constituents and their exact positions:

> Within the [railroad] yard, main-line through tracks are shown correctly placed, but other tracks are symbolized, preserving as much as possible the distinctive pattern presented by the yard.

> Spur tracks, sidings, switches, and storage tracks are mapped accurately as to length, but may be adjusted positionwise [that is, displaced] if the map scale and adjacent detail require it.

Although a thick book of dry rules no doubt takes some of the fun out of being a mapmaker, standardization lets the map user who learned to decode federal maps in one area transfer these skills with confidence to other parts of the country.

Cartographic guidelines have deep roots in military mapping. Founded in 1879, the USGS inherited mapping personnel from the four great post–Civil War surveys of the American West. Lead by Clarence King, F. V. Hayden, John Wesley Powell, and George M. Wheeler, these exploratory surveys addressed a variety of scientific, military, and economic goals and paved the way for the settlement of vast public lands in the little-known territories west of the 100th meridian. Surveyors and engineers typically had some military experience, and their mapping methods reflected techniques taught at West Point and used in the Union Army and the Corps of Engineers. As a result, topographic mapping in the new Geological Survey reflected military practice: a tradition fostered in later years by civil-service rules favoring war veterans and by executive orders requiring at least minimal coordination of civilian and military agencies with common interests.

Perhaps the most subtle example of military influence on civilian topographic maps is the green tint representing woodland.

Figure 10.2. Comparison of railway symbols on large-scale topographic maps of the United States (left) and Switzerland (right). (Excerpts enlarged to about 150 percent of original scale.)

Whether an area with trees is colored green reflects neither botanical nor ecological criteria but a tactical necessity. According to USGS topographic specifications, mappable woodland is "an area of normally dry land containing tree cover or brush that is potential tree cover"; that is, "the growth must be at least 6 feet (2 m) tall and dense enough to afford cover for troops." Traditional measurement units also influence cartographic specifications: for example, a small, isolated woods is shown only if it exceeds an acre in area, and a similar one-acre minimum controls the portrayal of clearings within a forest. European maps, by contrast, are typically affected by standards based on kilometers and hectares.

Although US military surveys of the nineteenth century learned (and borrowed) much from their European counterparts, large-scale US maps reflect uniquely American traditions in content and symbolization. A noteworthy example is our portrayal of railways with a thin line punctuated by short, uniformly spaced cross-ticks (fig. 10.2, left)—a cartographic icon so readily recognized by American map users that otherwise-complete map keys often omit it. Is this a natural symbol for railroads? Hardly: despite the vague suggestion of a steel rail running atop perpendicular wooden ties, the symbol's inherent graphic logic was surely not self-evident to the autocrats who established standards for topographic maps in Europe and elsewhere. Consequently, American tourists on their first trip across the Atlantic sometimes experience a jarring but usually short-lived cartographic culture shock when they encounter the heavy, black, unadorned Eurostyle railroad symbols with few visual cues for viewers who are unaware, for instance, that on Swiss maps (as in the right panel of figure 10.2) *Bhf.* is an abbreviation for *Bahnhof* (meaning "train station") and *Hst.* signifies a *Haltestelle* (literally a stopping place, or small station). As redundant or sole cues to meaning, abbreviations and feature names are valuable supplements to a purely geometric or pictorial graphic vocabulary.

Comparison of American and Swiss maps reveals numerous disparities in graphic conventions. As figure 10.3 illustrates, inherently ambiguous cartographic signs can have markedly different yet equally plausible meanings. To the Swiss, for instance, two thin parallel lines casing (enclosing) a thick dashed line represents a cog railway, which has steep grades ascended by a locomotive or power car equipped with a rotary cog engaging a rack of evenly spaced vertical teeth positioned between the rails. But to American map readers, a similar symbol with a dashed red fill quite logically represents a secondary highway, graphically intermediate between a primary road (shown by parallel black lines casing a solid red fill) and a light-duty road (similarly spaced black lines without the red fill). As another example, on Swiss maps a slight variation of the American railroad symbol portrays an aerial tramway, in which a car suspended from a moving cable supported by two or more towers ascends a steep slope or crosses deep gorges. In contrast, American maps take the two thick solid lines showing a double-track Swiss railway, spread them apart, and color them red to represent a limited-access dual highway. Whereas American mapmakers rely on various highway shields (interstate, US routes, state) to

Figure 10.3. Comparison of selected line symbols used on American and Swiss topographic maps.

differentiate more important routes, the Swiss show thoroughfares with a double-line symbol noticeably thicker on the bottom or right. Despite these radical differences, the American and Swiss cartographic vocabularies work equally well in their own milieu because map readers either learn the code or use a key.

Visitors to Switzerland are frequently amazed by this small, prosperous, and compulsively neutral nation's exquisitely detailed topographic maps. Close inspection suggests that the 1:25,000 topographic maps of the Bundesamt für Landestopografie are much richer and more thorough than their 1:24,000 American counterparts. Maybe, maybe not. A large part of the obvious difference is the faint gray hill shading that supplements elevation contours on Swiss maps. No doubt about it: because of cartographic enhancement, the portrayal of Swiss terrain is not only easier to interpret than that of American terrain but a significant aesthetic enhancement. Another prominent difference is the detailed Swiss portrayal of buildings in urban areas (fig. 10.2, right). In contrast, American topographic maps blanket built-up areas with a light red tint and show only landmark buildings, such as courthouses and churches. Whereas the Swiss seem preoccupied with geometry and terrain, American topographers are like effusive local boosters obsessed with pointing out churches, schools, and other features not differentiated on Swiss maps. In addition to mildly idiosyncratic symbolic vocabularies, American and Swiss mapmakers have developed different notions of what's worth showing. A comprehensive analysis of national mapping organizations would reveal surprising diversity in topographic vocabulary, with noteworthy differences between Canada and the United States, Austria and Germany, and Norway and Sweden. These differences have evolved somewhat like dialects, in response to necessity, isolation, and good ideas that catch on here but not there. Yet, unlike verbal language, cartographic symbols are more rigorously controlled through formal standards and specifications administered by institutions that resist change and controversy. That's why our topographic maps as well as their Swiss counterparts overlook toxic-waste dumps and evacuation zones around nuclear-power plants.

Cartographic Agendas

Whereas standards and specifications spell out much of the government mapmaker's duties, national mapping organizations typically pursue subtle strategies evident only through a careful reading of their products and activities. An exploration of these hidden cartographic agendas reveals a bureaucratic mentality eager to appear productive and sometimes willing to cut corners.

Well before the multicultural concept of hate speech, mapmakers were aware that place-names like "Niger [sic] Creek" (fig. 10.4, left)—even if misspelled—were racially offensive. Naively, if not innocently, late-nineteenth- and early-twentieth-century topographers striving for accuracy recorded these place-names as matter-of-fact representations of a geography authored by a politically incorrect generation of early Euro-American settlers, who might have intended no overt ill will in naming a feature Chink Creek, Dago Pass, Jap Gulch, or Nigger Lake. After World War II, as state legislatures and the US Congress banned racial discrimination in schools, public places, housing, and hiring, racial epithets—at least in official, public discourse—began to reflect more harshly

Figure 10.4. "Niger Creek" on a 1980 edition of the Bunker Hill, Indiana, 7.5-minute topographic map (left) was renamed "Negro Creek" on a 1994 revision (right). Had it been spelled with two *g*s, the racial slur might have been identified earlier when the Geological Survey searched its electronic geographic-names database for offensive toponyms.

on the speaker than on the target. And so the nation's large-scale maps became a time bomb of outrage and embarrassment, with cartographic bureaucrats reluctant to act because of their own rigid rules.

As with everything else on topographic maps, naming and renaming physical features had become a formal process, overseen in this case by the Domestic Names Committee of the US Board on Geographic Names, which spends much of its time reviewing recommendations from its state-level counterparts. Despite official channels being available to anyone offended by a name on a map, getting rid of one name usually meant replacing it with another. Sounds simple, but documenting an acceptable replacement name can be a contentious, time-consuming process. Few offensive names disappeared through direct appeal to the board.

Eventually the federal government did act, but only at the request of the Department of the Interior, which includes the Geological Survey. Separate directives called for changing all occurrences of the two most worrisome slurs to comparatively neutral synonyms. In 1962—less than a decade before "black" became the preferred term for "African American"—Interior Secretary Stewart Udall ordered the substitution of "Negro" for "Nigger" whenever an offending map sheet was revised and reprinted. (Oddly, the example in figure 10.4, although corrected in a 1994 revision, had been overlooked in a 1980 edit.) And in 1967, a similar order replaced the pejorative "Jap" with the innocuous "Japanese." For whatever reason, Chinese and Hispanic Americans have yet to win blanket eradication of "Chink" and "Dago."

While liberals wring their hands over the mapmaker's do-it-by-the-book intransigence over racial slurs, fiscal conservatives might object with equal vigor to the existence of perfectly executed, largely blue maps of wholly inundated quadrangles in the middle of the Great Salt Lake. A case in point is the Rozel Point SW, Utah, 7.5-minute quadrangle, which derives its name from a land feature on the next row up, one sheet to the right. Except for the titles, grid lines, and marginal notations similar to those on maps of Fresno and Kalamazoo, the 1:24,000 Rozel Point SW sheet is a featureless light-blue rectangle adorned only by a note at the center

ELEVATION 4200

providing the approximate elevation of the water surface above sea level but revealing nothing about the water's depth. A round number is appropriate because the Great Salt Lake, like other water bodies in the Great Basin, rises and falls in response to rain, snowmelt, and evaporation. Nice to know, though, that a topographic map is ready for revision should the lake dry up and reveal some real topography.

Two other, more plausible explanations exist, neither particularly reassuring. Taking the high road, mapmakers can argue that featureless quadrangles are quadrangles nonetheless, and that completeness demands maps for all inland quadrangles, regardless of content. After all, who knows when an obsessive limnologist might want to decorate an atrium with a collage of topographic maps showing the entire Great Salt Lake. And at a more venal level, a production-conscious supervisor might cleverly enhance the annual report by adding at low cost to the tally of "quadrangles completed." Because the first motive provides a plausible cover for the second, a joint explanation is likely.

If this portrait of a cost-conscious federal bureaucracy addicted to completeness seems far-fetched, consider the "provisional maps" introduced in the early 1980s. Pressured by the White House to cut costs yet eager to complete the 7.5-minute series, Geological Survey managers reviewed their multistep topographic mapping production process. An evaluation of different strategies suggested that the average number of person-hours per quadrangle could be trimmed from 745 to 573 by increasing the amount of preliminary research and postsurvey compilation while cutting back on field survey, graphic artwork, and editing. Though they might not look as nice, the new maps would still meet national map accuracy standards. And the definition of "provisional" as "prepared beforehand or temporary, to be followed by something permanent" promised the few irate map connoisseurs a more aesthetically satisfactory product at an unspecified later date.

How aesthetically deficient are these economy-grade maps? Let the excerpt in figure 10.5 speak for itself. Although cartographic beauty is forever in the eye of the beholder, scratching on the negative with a sharp instrument to make faint, jerky labels like "Boat Ramp" and "Episcopal Ch. Spire" was a Neanderthal approach to what the Geological Survey calls "map finishing." Although labels identifying minor landmarks, locally important facilities, and survey markers (like the 4-foot tide-level benchmark coded "BM 4 – Tidal") are visually recessive, scratch lettering reflects an appalling disdain for style, grace, and legibility. And as a consequence of less editing and difficult corrections, provisional maps are more vulnerable than conventional topographic maps to spelling errors.

Ironically, about the time its drafters regressed to scribble, the Geological Survey tarted up the 7.5-minute provisional maps and a new series of 1:100,000 maps with a distinctive typeface named Souvenir. For an example of this distinctiveness, compare "Stoddartsville" in figure 10.6 with the more traditionally rendered

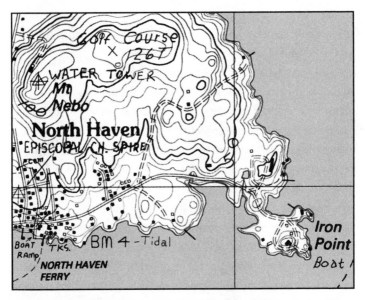

Figure 10.5. The US Geological Survey expedited completion of its 7.5-minute, 1:24,000 series by scratching crude labels on photographic negatives of "provisional-edition" maps, such as the North Haven East, Maine, sheet (enlarged to about 150 percent of original scale).

"McGrawsville" in figure 10.4. Note, in particular, the stubby serifs (short strokes at the ends of letters) and distinguishing curved diagonal strokes in the lowercase letter v as well as in capital letters such as K, U, and Z. Advertising art directors liked this warm, contemporary look, and Pizza Hut, a national fast-food chain, adopted Souvenir for its ads and menus.

What Geological Survey managers were up to no one can fathom with certainty. Perhaps they wanted to give their product a contemporary, user-friendly appearance, thereby promoting sales at the expense of the topographic map's traditional aura of authority and power. Maybe a top-level manager saw Souvenir, liked it, and encouraged subordinates to use it. Possibly a design consultant with a background in advertising recommended it. In a more cynical interpretation, though, USGS officials might have recognized the power of bold, thick labels like "North Haven" in figure 10.5 to draw attention from the emaciated lettering on their provisional

Figure 10.6. Many place-name labels on 1:100,000 series maps are set in Souvenir, as illustrated by the township (Buck), village (Stoddartsville), and county names on this excerpt (enlarged to about 150 percent of original scale) from the 1986 planimetric (without contours) version of the Scranton, Pennsylvania, sheet.

maps. Whether by design or happenstance, this decoy effect is an appropriate form of face-saving for fast-food cartography.

Sacrificing appearance to save money can be serendipitous, as photorevised 7.5-minute topographic maps clearly demonstrate. To cope with keeping fifty-four thousand topographic quadrangle maps reasonably up-to-date, the Geological Survey adopted a form of triage. Every five to ten years, on average, cartographers compare the most recent edition of a map sheet with new aerial photography in order to decide whether sufficient change has occurred to warrant revision and, if so, whether to draw an entirely new map or to produce a photorevised edition by merely adding new features such as roads and buildings in purple ink. The results look garish (especially if you don't like purple) as well as inconsistent (for instance, when contour lines around new limited-access dual highways are not revised to show cuts and fills), but the obvious advantage of more timely maps for many more quadrangles than otherwise possible outweighs these aesthetic shortcomings.

Figure 10.7. Excerpts from various editions of the 1:250,000 Harrisburg, Pennsylvania, map chronicle the puzzling attempt to abandon a nonexistent, erroneously insert-ed railroad running northeast from State College. (Maps reduced from 1:250,000 to approximately 1:330,000.)

And, however inelegant, purple symbols usefully highlight areas and type of change as well as warn map users that static maps almost always lag behind the dynamic landscape they pretend to portray. All maps lie, but few are as frank as photorevised topographic sheets.

Committed to geometric accuracy and geographic completeness, federal topographers seldom make outrageous mistakes, but when they do, the images are memorable. The most outrageous example I know involves an equally outrageous cover-up. Figure 10.7 shows a nonexistent railway running northeast from State College, Pennsylvania, to Dale Summit and then turning north along state highway 545. This cartographic fiction appeared on the 1:250,000 Harrisburg sheet in a 1957 edition, compiled by Army Map Service personnel, who also worked on the 1:250,000 series. Although the compiler might have resurrected the short abandoned branch line from State College south to Pine Grove Mills from air photos or an old map, no railway ever ran northeast of town, through a business district and an established residential neighborhood and across moderately steep terrain, which would have required at least one noteworthy trestle or an enormous fill—having lived there, I know. But that's only half the story. A 1965 edition with "limited revision by the U.S. Geological Survey" caught the error, but instead of removing the faulty railway symbol altogether, the cartographic artist substituted an abandoned railroad, represented by an interrupted, dashed version of the traditional cross-ticked railway symbol. By 1969, the federal topography factory finally got it right and dropped the fictitious feature once and for all.

Although gratuitous symbols usually reflect blunders, deletions typically follow careful deliberation of official feature standards. In the 1980s, for instance, when thoughtless souvenir hunters and other looters became especially troublesome at historic sites in the southwest, the National Park Service asked the Geological Survey, the Automobile Club of Southern California, and other map publishers to remove ruins and other archaeological sites from their

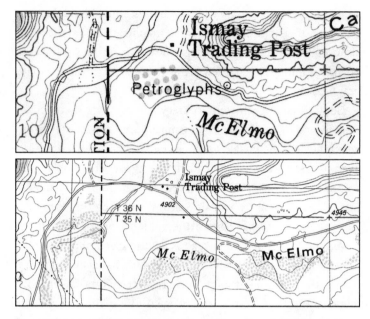

Figure 10.8. A late 1950s topographic map (above, enlarged from 1:62,500 to about 1:24,000) pointed out the petroglyphs just southeast of the Ismay Trading Post, but more recent 7.5-minute maps (below) omit this and numerous other archaeological sites.

maps and guidebooks. In complying, USGS mapmakers eliminated the petroglyphs (rock carvings) identified on the 1958 15-minute topographic sheet covering Ismay, Colorado (fig. 10.8, above), from the more detailed 7.5-minute map issued in 1985 (fig. 10.8, below). Most cultural sites attractive to treasure hunters will no doubt disappear from maps as revision progresses, however slowly, but with blatant clues readily available in map archives, the policy of suppressing archaeological features seems a bit late. Although the cartographic contribution to cultural looting is impossible to assess, accuracy and completeness clearly can have unintended consequences.

National mapping organizations are easy to criticize for much the same reason maps must lie: the enormous number of choices in selecting features, assigning symbols, and setting the scale of a map series. There's the added responsibility, though, of keeping tens of thousands of large-scale maps current through periodic review and revision. Because of limited resources, even a range of map series cannot satisfy the diverse requirements of map users. And while agencies cope with political pressures for greater cost recovery, more effective federal-state-local coordination, and increased privatization, technological advances create ever more promising yet costly choices as well as a growing need to reassess priorities and revise standards. To be adequately informed, the map user must be at least vaguely aware of how cartographic bureaucracies work, what they value, and how values and biases affect their products.

Online Alternatives

Although the Geological Survey's website now offers downloadable digital versions of its topographic maps, these static products are less versatile in many ways than the dynamic online maps provided by Google and its competitors. Although brown elevation contours have long been the hallmark of the USGS quadrangle map, the term *topography*, which means "place description," is broader in scope than landforms and vertical position. While online topographic maps from private firms pay less attention to elevation, they compensate nicely by including street names and by labeling private buildings whose owners or tenants pay for the privilege. Indeed, advertising, whether on the map proper or in an adjacent panel, is online topographic mapping's principal support. By contrast, national mapping agencies still provide the scientific geographic data required for environmental protection, emergency management, national defense, and the efficient management of federal lands.

This public/private relationship is a partnership of sort insofar as the private sector has built on the federal cartographic framework, including the vast database of officially sanctioned place- and feature names, to serve an important niche once met by commercial publishers of road maps and indexed street maps. The relationship is more symbiotic than economic insofar as online mapping is an outgrowth of the Census Bureau's development of a digital street map, originally intended to relate household addresses to tabulation areas called *census tracts*. The private sector then added address-locator and route-following services. Our current mapping culture thus reflects two traditions: free access to publicly funded spatial data and advertiser support of private-sector information services—the latter tradition pioneered by newspapers and the broadcast media. Much of the technology that supports online mapping can be traced to federal support for digital cartography, particularly for national defense.

Our new private-sector national map is revolutionary in many ways, most notably in the rapid updating of information, the seamlessness of spatial data no longer hamstrung by the network of quadrangle boundaries, and the user's ability to zoom in and out to more- or less-detailed views. No less revolutionary is the user's ability to toggle between a more or less conventional topographic map and a detailed overhead image of vegetation, structures, and swimming pools. What's more, the interactive street view lets the user gaze onto front porches and into side yards. Although American culture seems willing to accept this mild form of intrusion— we have constitutional protections against warrantless electronic penetration of roofs and walls—several European nations have restricted remote viewing of images captured with mobile cameras in streets and other public spaces. Concern for privacy, as a manifestation of culture and politics, is setting a limit to topographic detail.

Data Maps:
A Thicket of Thorny Choices

A single set of numerical data can yield markedly dissimilar maps. By manipulating breaks between categories of data to be shaded on a choropleth map, for instance, a mapmaker can often create two distinctly different spatial patterns. A single map is thus just one of many maps that might be prepared from the same information, and the map author who fails to look carefully at the data and explore cartographic alternatives easily overlooks interesting spatial trends or regional groupings.

Wary map users must watch out for statistical maps carefully contrived to prove the points of self-promoting scientists, manipulating politicians, misleading advertisers, and other propagandists. Meanwhile, this is an area in which the widespread use of mapping software has made unintentional cartographic self-deception inevitable. How many software users know that using area-shading symbols with magnitude data produces misleading maps, or that size differences between areal units such as counties and census tracts can radically distort map comparisons?

This chapter uses several simple hypothetical examples featuring a fictional electronic device we'll call a "gizmo" to examine the effects of areal aggregation and data classification on mapped patterns. Anyone interested in public-policy analysis, marketing, social science, or disease control needs to know how maps based on numbers can yield useful information as well as flagrant distortions.

Aggregation, Homogeneity, and Areal Units

Most quantitative maps display data collected for areas such as counties, states, and countries. When displayed on a map, present-

Number of Gizmos

1,000	100	50	100	50	100	50
200	100	200	100	200	100	200
100	200	100	4,000	100	200	100
200	400	200	400	200	400	3,000

Number of Households

2,000	200	100	200	100	200	100
200	100	200	100	200	100	200
100	200	100	4,000	100	200	100
100	200	100	200	100	200	1,500

Gizmos per Household

0.5	0.5	0.5	0.5	0.5	0.5	0.5
1.0	1.0	1.0	1.0	1.0	1.0	1.0
1.0	1.0	1.0	1.0	1.0	1.0	1.0
2.0	2.0	2.0	2.0	2.0	2.0	2.0

Figure 11.1. Town-unit number tables showing number of gizmos (top left), number of households (top right), and average number of gizmos per household (bottom) for twenty-eight hypothetical towns.

Number of Gizmos

2,300	5,700	4,150

Number of Households

3,100	5,500	2,600

Gizmos per Households

0.74	1.04	1.60

Figure 11.2. County-unit number tables of number of gizmos (left), number of households (middle), and average number of gizmos per household (right) for a three-county aggregation of the twenty-eight hypothetical towns in figure 11.1.

ed on a statistical plot, or analyzed using correlation coefficients or other measures, geographic data produce results that reflect the type of areal unit. Because different areal aggregations of the data might yield substantially different patterns or relationships, the analyst should qualify any description or interpretation by stating the type of geographic unit used. Noting that values generally increase from north to south "at the county-unit level" warns the reader (and the mapmaker as well!) that a different trend might arise with state-level data, for instance.

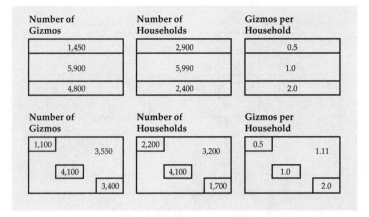

Figure 11.3. County-unit number tables based on other aggregations of the twenty-eight towns into counties.

Areal aggregation can have a striking effect on the mapped patterns of rates and ratios. A ratio such as the average number of gizmos per household might, for example, produce radically different maps when the data are aggregated separately by counties and by the towns that make up these counties. The three town-level maps in figure 11.1 are spatially ordered number tables, without graphic symbols, so that we can see how rate calculations depend on what boundaries are used and how they are drawn. The upper left-hand map shows the number of gizmos in each of twenty-eight towns, the upper right-hand map represents the number of households, and the lower map portrays the gizmo-ownership rate. Note the straightforward top-to-bottom pattern of the rates: low in the upper tier of towns, average in the two middle tiers, and high in the lower tier. Note also that three towns in the upper left, lower right, and just below the center of the region have relatively high numbers of households. These variations in household density underlie the markedly different left-to-right trend in gizmo-ownership rates in figure 11.2, based on the same data aggregated by county.

Spatial pattern at the town-unit level of aggregation depends on how somewhat arbitrary political boundaries group towns into counties. Figure 11.3 uses two additional aggregations of these

Number of Gizmos

190	285	200	350	350	210	890
455	450	1,085	960	895	520	1,260
355	315	525	480	595	360	700
130	120	80	100	80	110	100

Number of Households

100	150	100	200	100	50	100
350	300	700	600	500	200	300
500	450	700	600	700	400	500
650	600	400	500	400	550	500

Gizmos per Household

1.90	1.90	2.00	1.75	3.50	4.20	8.90
1.30	1.50	1.55	1.60	1.79	2.60	4.20
0.71	0.70	0.75	0.80	0.85	0.90	1.40
0.20	0.20	0.20	0.20	0.20	0.20	0.20

Figure 11.4. Patterns of the number of gizmos, the number of households, and the gizmo-ownership rate radically different from those in figure 11.1 could yield county-unit patterns identical to those in figure 11.2.

twenty-eight towns to demonstrate the possible effect of historical accident. The upper row of maps shows an alternative aggregation of towns into three horizontal counties that reflect the town-level top-to-bottom trend. In contrast, the lower series of maps shows an equally plausible aggregation into four counties, three based on the concentrations of households and one comprising the balance of the region. The gizmo-ownership map for this lower set isolates what might be more urban counties from a single much larger, more rural county with an average of slightly more than one gizmo per household. Graytone area symbols would yield very different choropleth maps for the three sets of rates shown in the right-hand maps of figures 11.2 and 11.3.

Another example illustrates how areal aggregation can affect geographic pattern. Whereas figure 11.3 demonstrates that different aggregations of towns into counties can yield markedly different county-level patterns, figure 11.4 illustrates how a single aggregation can produce the same county-level pattern from markedly different town-level patterns. Note that the town-level maps in figure 11.4 reflect a pattern of gizmo-ownership rates very different from that in figure 11.1. Note in particular the progression of rates from a tier of low-ownership towns across the bottom of the region to a peak of much higher rates at the upper right. Yet when aggregated according to the county boundaries in figure 11.2, these data will yield similar county-unit rates. Comparing this trio of spatial number tables with those in figure 11.1 demonstrates the importance of stating clearly the data units used and of not assuming that a trend apparent at one level of aggregation exists at other levels as well.

The counties in these examples obviously are not homogeneous. But can we assume homogeneity even within the towns? What spatial variations in the distribution and density of these 11,200 households lie hidden in the network of town boundaries? Figure 11.5 presents one of many plausible point patterns that could produce the aggregated town-level counts and rates in figure 11.1. Three types of point symbols represent groups of ten, one hundred, and five hundred households. Each symbol represents a group of households owning an average of zero, one, or two gizmos. The small, ten-household symbols represent rural residences, which might lack gizmos because of a lack of connectivity, less spare time, or a low opinion of digital gadgetry. Because of rough terrain, swamps, park- or forestland, and undeveloped federal land, large parts of the region are uninhabited. Of the six large villages, with four hundred or more households, two have two-gizmo households on the average, two have one-gizmo households, and two have gizmo-free households. Although figure 11.5 contains elements of both the top-to-bottom town-level trend in figure 11.1 and the left-to-right county-unit trend in figure 11.2, its pattern of gizmo ownership is more similar to the lower right of figure 11.3, where county boundaries segregate three large population clusters from the balance of the region. Yet even here the differences are striking, again demonstrating how the configuration of areal units can hide interesting spatial detail and present a biased view of a variable's geography.

158

CHAPTER 11

Figure 11.5. Detailed map of gizmo ownership for villages and rural households illustrates one possible spatial structure that could yield the town-unit and county-unit maps in figures 11.1 and 11.2.

Aggregation's effects become even more serious if the careless analyst or naive reader leaps from a pattern based on areal units to conclusions based on individual households. Consider, for instance, the large village toward the lower right-hand corner of figure 11.5. The average gizmo-ownership rate here of 2.0 need not mean that each of the village's 1,700 houses has two gizmos. Some households might have none while others might have three or four or five. One or two residents might even be compulsive collectors—hoarders masquerading as hobbyists—so that more than half the homes have one or none.

If households collecting old gizmos seems far-fetched, consider average household income, an index used frequently by social scientists and marketing analysts. Because of one or two innovative, unscrupulously manipulative, or otherwise successful residents, a small village might have an enormous *mean* household income. More of a statistical quirk than a realistic reflection of overall local prosperity, this high average income might mask the employment of most villagers as household servants, gardeners, or security guards. Because nondisclosure rules prohibit a more precise publication of individual incomes, aggregated census data are the most refined information available. They provide an average for the place but say little about individual residents.

Are areally aggregated data bad? Surely not. In many cases, particularly in public-policy analysis, towns and counties are the truly relevant units for which state and federal governments allocate funds and measure performance. And even more highly aggregated data can be useful, for instance, when governors and senators want to compare their states with the other forty-nine. Local officials and social scientists concerned with differences between neighborhoods readily acknowledge the value of geographic aggregation. Moreover, nondisclosure regulations, which are needed to ensure cooperation with censuses and surveys, require aggregation, and areally aggregated data are better than no data at all. Thus persons who depend on local-area data encourage the Census Bureau to modify boundaries to preserve the homogeneity of census tracts and other reporting areas. And when tract data are not adequate, they sometimes pay for new aggregations of the data to more meaningful areal units.

What else can the conscientious analyst do? Very little aside from the obvious: know the area and the data, experiment with data for a variety of levels of aggregation, and carefully qualify all conclusions.

And what should the skeptical map user do? Look for and compare maps with different levels of detail, and be wary of cartographic manipulators who choose the level of aggregation that best proves their point.

Aggregation, Classification, and Outliers

Choropleth mapping further aggregates the data by grouping all areas with a range of data values into a single category represented by a single symbol. This type of aggregation addresses the difficulty of displaying more than six or seven visually distinct colors or graytones in a consistent light-to-dark sequence. Often the mapmaker prefers only four or five categories, especially when the area symbols available do not afford an unambiguous graded series. (For aesthetic reasons or to avoid confusion with interior

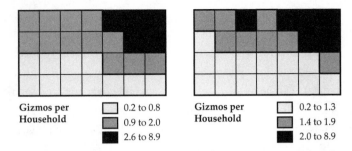

Gizmos per Household
- ☐ 0.2 to 0.8
- ▨ 0.9 to 2.0
- ■ 2.6 to 8.9

Gizmos per Household
- ☐ 0.2 to 1.3
- ▨ 1.4 to 1.9
- ■ 2.0 to 8.9

Figure 11.6. Different sets of categories yield different three-category choropleth maps for the data in figure 11.4.

lakes or areas without data, opaque white and solid black are not good graytone symbols for choropleth maps.)

But classification introduces the risk of a mapped pattern that distorts spatial trends. Arbitrary selection of breaks between categories might mask a clear, coherent trend with a needlessly fragmented map or oversimplify a meaningfully intricate pattern with an excessively smoothed view. Figure 11.6 illustrates the influence of class breaks on the appearance of choropleth maps of the town-level gizmo-ownership rates in figure 11.4. Note that the map on the left presents a clear, straightforward, readily remembered upward trend toward a peak at the upper right of the region, whereas the map at the right offers a more fractured view of the same data.

Classification raises many questions. Which map, if either, is right? Or if "right" sounds too dogmatic, which provides a better representation of the data? Don't both maps hide much variation in the broad third category, represented by the darkest symbol? Shouldn't the seven towns with rates of 0.2 occupy a category by themselves? Is a difference of, say, 0.1 at the lower end of the overall range of data values more important than a similar difference at the upper end? Can a three-class map provide even a remotely adequate solution?

These questions are vital not only to map users but also to map authors, particularly those using graphics software but untrained

Figure 11.7. Two common classing schemes used as "defaults" by choropleth mapping software yield radically different four-category patterns for the data in figure 11.4.

in cartography. Software applications usually provide a few options for "automatic" classification, and naive mapmakers often settle for one of the easier options. Sometimes the software even provides a map instantly, without offering a choice of classification strategies. Called a *default option*, this automatic choice of class breaks is a marketing ploy that gives the hesitant prospective purchaser an immediate success.

But does the default give you a good map? Figure 11.7 shows four-category mapped patterns produced by two common default classing options for the same town-level gizmo-ownership data used in figure 11.6. The *equal-intervals* scheme, on the left, divides the range (8.7) between the lowest and highest data values (from 0.2 to 8.9) into four equal parts (each spanning 2.175 units). Note, though, that this classification assigns most of the region to a single category and that the third category (from 4.6 to 6.7) is empty. Of possible use when data values are uniformly distributed across the range, the only consistent asset of equal-interval classification is ease of calculation.

By contrast, the *quartile* scheme, on the right, ranks the data values and then divides them so that all categories have the same number of areal units. Of course, only an approximately equal balance is possible when the number of areas is not a multiple of four or when

Figure 11.8. A continuous-tone, non-classed choropleth map for the data in figure 11.4.

a tie thwarts an equal allocation (as occurs here at the upper left, where the highest category receives both of the towns with rates of 1.9). Although the map pattern is more visually balanced, the upper category is broad and highly heterogeneous, and the break between the second and third categories falls between two very close values (1.3 and 1.4). Yet the map based on these four quartile categories does have meaning for the user interested in the locations of towns in the highest and lowest quarters of the data values. Called *quintiles* for five categories and *quantiles* more generally, this rank-and-balance approach can accommodate any number of classes.

Some mapping applications offer the option of a "no-class" or "class-less" choropleth map, on which each unique data value (perhaps up to fifty of them) receives a unique graytone. In principle this might seem a good way to sidestep the need to set class breaks. But as figure 11.8 illustrates, the graytones might not form a well-ordered series, and the map key is either abbreviated or cumbersome. Moreover, assigning each unique value its own category can

Gizmos per Household

Figure 11.9. Number line for the town-level gizmo-ownership rates in figure 11.4.

destroy a clear, easily remembered picture of a strong, meaningful spatial trend. This ideal solution might not be so ideal after all.

Eschewing defaults and panaceas, the astute map author begins by asking two basic questions: How are the data distributed throughout their range? And what, if any, class breaks might have particular meaning to the map user? The answer to this second question depends on the data and on whether the map author deems useful a comparison with the national or regional average. On state-level maps, for instance, a break at the United States average would allow governors and senators to compare their constituents' or their own performance with that of the rest of the nation. Of course, the map key would have to identify this break to make it truly meaningful.

After addressing the question of meaningful breaks, the conscientious map author might then plot a *number line* similar to that in figure 11.9. A horizontal scale with tick marks and labels represents the range of the data. Each dot represents a data value, and identical values plot at the same position along the scale, one above the other. The resulting graph readily reveals natural breaks, if any occur, and distinct clusters of homogeneous data values, which the classification ought not subdivide. Number lines allow the map author to visualize the distribution of data values and to choose an appropriate number of categories and appropriate positions for class breaks. Computer algorithms can also search the data distribution for an optimal set of breaks, but in many cases the computer-determined optimum is not significantly better

Gizmos per
Household

☐ 0.2
☐ 0.7 to 0.9
☐ 1.3 to 2.0
☐ 2.6 to 4.2
☐ 8.9

Figure 11.10. Choropleth map based on the number line in figure 11.9 and the character of the data.

than a visually identified suboptimal grouping. Rounded breaks and a more balanced allocation of places among categories can be important secondary factors in choropleth mapping.

Extremely high or extremely low values isolated from the rest of the distribution can confound both human cartographers and sophisticated mapping software. Should these *outliers* be grouped with markedly more homogeneous clusters higher or lower on the number line? Should each be accorded its own category? Can two or three widely separated data values at either end of the distribution be grouped into a single, highly heterogeneous category? Or should each outlier be treated as its own category, with its own symbol, at the risk of reducing graphic differentiation between area symbols? Or might the map author treat outliers as outcasts— errors or deviants that "don't belong"—and either omit them or give them a special symbol?

No simple, standard solution addresses all outliers. The map author should know the data, know whether these deviant values are real or improbable, and know whether a large difference between outliers really matters. Also important is the relation of outliers to the theme of the map and the interests of map users. For the gizmo-ownership data in figure 11.9, an average of 8.9 gizmos per household surely is not only exceptional but probably significantly higher than its neighboring values at 4.2. If not an error, it deserves special treatment in a category of its own. The next four lower values, 4.2 (twice), 3.5, and 2.6, might then constitute

a single category; all are above the more plausible rate of 2.0, and
yet 4.2 gizmos per household is not improbable, especially in an
affluent area.

Other breaks seem warranted between 0.9 and 1.3, a gap that
includes the inherently meaningful rate of one gizmo per household, and between 0.2 and 0.7, to separate the seven techno-
phobic towns at the lower end of the distribution. The resulting
five-category map in figure 11.10 provides not only an honest,
meaningful representation of the data values and their statistical
distribution, but a straightforward portrayal of the spatial trend
as well. An arbitrary classification, such as a computer program's
default categories, is unlikely to do as well, even with six or more
categories.

Classification, Correlation, and Visual Perception

Choropleth maps readily distort geographic relationships between
two distributions. Hastily selected or deliberately manipulated
categories can diminish the visual similarity of two essentially
identical trends or impose an apparent similarity between two
very different patterns.

Consider as a case in point figure 11.11, a spatial-data table and
number line for the mean number of children per household,
which has a strong town-level relationship to gizmo ownership.
Although the range of data values is not as broad for this index of
family size, the highest values are at the upper right and the lowest
values occur across the bottom of the region. Towns toward the
right and toward the top of the region generally have more children
in the home than do towns toward the bottom or left edge of the
map. That the pair of maps in figure 11.12 shows identical spatial
patterns for children and gizmos is thus not surprising.

Statistical analysts commonly depict correlation with a two-
dimensional scatterplot, with data values for one variable measured
along the vertical axis and those for the other scaled along the horizontal axis. A dot represents each place, and the density and orien-

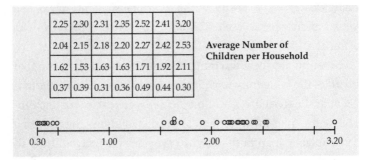

Figure 11.11. Spatial-data table and number line for average number of children per household.

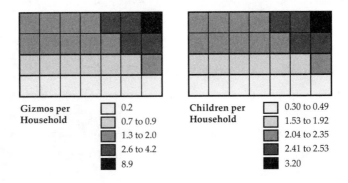

Figure 11.12. Choropleth maps with identical patterns for gizmo-ownership rate and average number of children per household.

tation of the point cloud indicates the strength and direction of the correlation. Figure 11.13 is a pair of scatterplots, both showing the strong positive association between the household rates for children and gizmos. The perpendicular lines extending from the scales of the left-hand scatterplot into the scatter of points represent the class breaks in figure 11.12. These two sets of four lines each divide the scatterplot into an irregular five-by-five grid. Because all dots on the left-hand scatterplot lie within one of the five diagonal cells, the two five-category maps in figure 11.12 have identical patterns, enhancing the impression of a strong correlation.

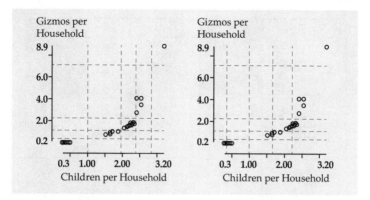

Figure 11.13. Scatterplots for the town-level gizmo-ownership rate and average number of children per household. Additional lines on the left-hand scatterplot represent class breaks for the pair of maps in figure 11.12. Additional lines on the right-hand scatterplot show breaks used in figure 11.14.

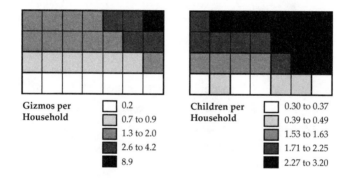

Figure 11.14. Distinctly different choropleth maps suggest minimal correlation between gizmo ownership and family size.

Figure 11.13's right-hand scatterplot adds some cartographic skulduggery. As before, the perpendicular lines from the scales into the point cloud represent class breaks and form a five-by-five grid. But note that this configuration of breaks places all but four dots in an off-diagonal cell so that few towns will belong to the same category on both maps. Figure 11.14 demonstrates the resulting dissimilarity in map pattern and suggests a mediocre correlation at best. Similar tactics might make a weak relationship appear strong,

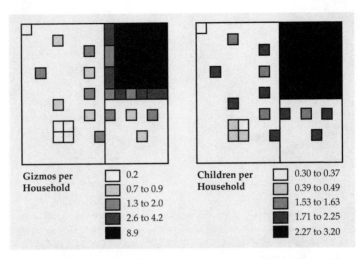

Figure 11.15. Similarity between large areas can distort visual estimates of correlation by masking significant dissimilarity between small areas. Numerical data and mapping categories are identical to those for the more obviously dissimilar pair of maps in figure 11.14.

especially if the maps are identical for the highest category, with the darkest symbol. Indeed, the spatial correspondence of the darkest, most eye-catching symbols strongly influences judgments of map similarity by naive map users. Some will even regard as similar two maps with roughly equal amounts of the darkest symbol—even if the high areas are in different parts of the region! Different area symbols for the two maps and different numbers of categories are other ways of tricking the map user or deluding oneself.

Another visual distortion might lie in the base map the data are plotted on. Not all sets of areal units are as uniform and visually equivalent as the square towns in the preceding examples. Figure 11.15 demonstrates this point with a deceptively similar-looking pair of maps based on the numerical data and class breaks of the visually dissimilar maps in figure 11.14. These twenty-eight towns vary markedly in size, and similarity is high because the largest towns belong to the same category. Towns not in the same category on both maps are smaller and less visually influential.

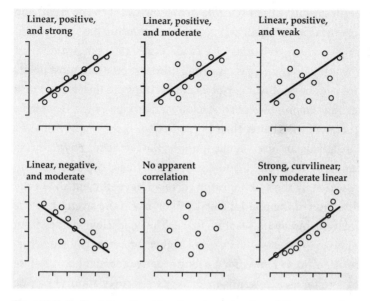

Figure 11.16. Scatterplots and trend lines for various types of correlation.

Although this example involving gizmo ownership is contrived, it is not atypical. Wards, census tracts, congressional districts, and other areal units designed to have similar populations often vary widely in area because of variations in population density. Disparities are even worse on county-unit maps, where populous metropolitan counties often are much smaller than rural counties with few inhabitants. The careful map user never judges numerical correlation by the similarity in map pattern alone and is especially cautious when some data areas are much bigger than others.

To avoid estimates of correlation biased by the size of areal units, the astute analyst will inspect the more egalitarian scatterplot, on which identical dots represent each area. As figure 11.16 illustrates, the density and orientation of the point cloud reflect the strength and direction of the correlation. If a straight line provides a good generalization of the point cloud, the correlation is called linear and the scatter of points around the line indicates the strength of the *linear correlation*. Positive relationships slope upward to the right, negative relationships slope downward to the right, and a point cloud without a discernible relationship has no apparent slope. Weak correlations have a wide, barely coherent scatter about the trend line, whereas for strong linear correlations most points are near or on the line. Not all correlations are linear, though; a strong *curvilinear correlation* has a marked curved trend, which a curved line fits better than a straight line.

Statisticians use a single number, the *correlation coefficient*, to measure the strength and direction of a linear correlation. Represented by the symbol r, the correlation coefficient shows the direction of the relationship by its sign and the strength of the relationship by its absolute value. The coefficient ranges from +1.00 to −1.00; r would be 0.9 or higher for a strong positive correlation, −0.9 or lower for a strong negative correlation, and close to zero for an indeterminate or very weak correlation. (As a rule of thumb, squaring r yields the proportion of one variable's variation accounted for by the other variable. Thus, if r is −0.6, the correlation is negative and one variable might be said to "explain" 36 percent of the other variable. A correlation coefficient measures only association, not causation, which depends on logic and supporting evidence.)

Maps, scatterplots, and correlation coefficients are complementary, and the analyst interested in correlation relies on all three. The correlation coefficient, which provides a concise comparison for a pair of variables, measures only linear correlation. Yet a scatterplot quickly reveals a strong curvilinear relationship, with a mediocre value of r. Scatterplots also show outliers, which can greatly bias the calculation of r. But reliance on visual estimation makes scatterplots poor for comparing strengths of relationships. Moreover, scatterplots and correlation coefficients tell us nothing about the locations of places, whereas maps, which present spatial trends, can offer unreliable estimates of correlation.

Maps also show a different kind of correlation, a *geographic correlation* distinct from the statistical correlation of the scatterplot and the correlation coefficient. Statistical correlation is aspatial

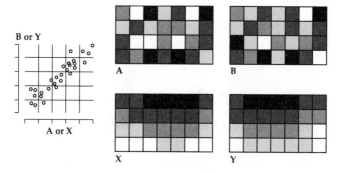

Figure 11.17. Two pairs of variables with identical scatterplots, correlation coefficients (r = .93), and class breaks, yet distinctly different map patterns.

and reveals nothing about spatial trends. Figure 11.17 demonstrates this difference with two map pairs distinct in spatial pattern yet identical in scatterplot and correlation coefficient. Variables A and B, which share a comparatively chaotic, fragmented pattern, clearly differ in geographic correlation from variables X and Y, which have a distinct common trend with higher values toward the top of the region and lower values toward the bottom. Although not identical, the maps for X and Y suggest the influence of a third, underlying geographic factor, such as latitude, ethnicity, soil fertility, or proximity to a major source of pollution. Despite the problems posed by areal aggregation, the analyst of geographic data who explores correlation without also checking for spatial pattern is either ignorant, careless, or callous. And the nonskeptical reader is easily misled.

Whether expressed numerically or with maps, correlations based on spatially aggregated data are vulnerable to the *ecological fallacy*, whereby a relationship demonstrated for one level of areal aggregation—say, with county units—is presumed to hold for other aggregations (such as states) as well as for individuals. (Regardless of their size, areal units are considered *ecological* units, rather than individuals.) For example, a finding that areas above average in number of years of education tend to be above average in income does not mean that people with master's degrees are necessarily well paid—graduate students pursuing a doctorate are a case in point.

Places, Time, and Small Numbers

Areal data can yield particularly questionable patterns when choropleth maps show rates based on infrequent events, such as deaths from a rare type of cancer. Yet disease maps based on small numbers are a common tool of the epidemiologist, who uses mapping to explore the possible effects on human health of radon-rich soils, incinerators, chemical-waste dumps, and drinking water supplied through lead pipes. But one question arises whenever the map shows a trend or cluster: Is the pattern real?

The problem is one of small numbers. Pandemics are rare, and seldom is the association between disease and an environmental cause so overwhelming that the link is easily identified and unchallenged. Clusters of deaths or diagnosed cases usually are few and unspectacularly small, perhaps no more than three deaths in a town or two in the same neighborhood. Epidemiologists map these cases both as points, to get a sense of patterning, and by areal units, to adjust for spatial differences in the number of people at risk. After all, an area with half the region's cases is not remarkable if it has half the region's population. But what is the significance of a small area with two or three cases and a rate several times above the national or regional rate? Could this pattern have arisen by chance? Would one or two fewer cases make the area no longer a "hot spot"? If one more case were to occur elsewhere, would this other area also have a high rate? To what extent does the pattern of high rates reflect arbitrary boundaries, drawn in the last century to promote efficient government or decades ago to expedite delivery of mail? Might another partitioning of the region yield a markedly different pattern? Might another level of aggregation—larger units or smaller units—alter the pattern? Is the mapping method inflating the significance of some clusters? And is it possibly hiding others?

Figure 11.18. A reconstruction of John Snow's famous dot map of cholera (above) and three choropleth maps (below) produced by different areal aggregations of this part of London.

Consider, for example, the maps in figure 11.18. At the top is a reconstruction of John Snow's famous map showing cholera deaths clustered around the Broad Street pump. A physician working in London during the cholera epidemic of 1854, Snow suspected drinking water as the source of infection. At that time homes did not have running water, and people carried buckets from a nearby pump. According to legend, Snow's map confirmed the waterborne transmission of cholera, and when authorities removed the pump's handle, new cases in this part of the city plummeted. Truth be told, the epidemic had run its course, and Snow made his map months later when he revised his book on cholera.

But what might have happened had Snow not worked with point data? The three maps at the bottom of figure 11.18 show how various schemes of areal aggregation might have diluted the Broad Street cluster. If addresses are available, as on most death certificates, aggregation to census tracts or other areal units larger than the city block increases the risk of missing intense, highly local clusters.

Aggregation involves not only areal units but also time, disease classification, and demography. One solution to the question of significance is to get more data by collecting information over a longer time span. Adding together several years of data, or even several decades, dampens the effect of chance occurrences but risks involving a wider range of causal agents. Aggregation over time might, for instance, mask important temporal trends, dilute the impact of new or abated environmental contaminants, or incorporate difficult-to-measure effects of population mobility. Likewise, combining several disease categories or the mortalities of diverse demographic groups promotes stability and significance by increasing the number of cases and broadening the set of causes.

Clearly one map is not sufficient, although one good map can signal the need for a more detailed investigation. It is then up to a variety of scientific researchers to explore further the effects of geography and environment by examining employment and residential histories, characteristics of residence and neighborhood, and hereditary factors; by carefully studying maps at various levels of spatial, temporal, and demographic aggregation; through computer simulation to test the stability of known clusters; through automated pattern recognition to identify new clusters; and through related clinical and laboratory studies. Although maps can indeed lie, they can also hold vital clues for the medical detective.

Indexes, Rates, and Rates of Change

Another danger of one-map solutions is a set of measurements that presents an unduly positive or negative view. Often the map author has a single theme in mind and has several variables to choose from. Usually some variables are markedly more optimistic in tone or pattern than others, and the name of the index can cast a favorable or unfavorable impression in the map title. "Labor Force Participation," for instance, sounds optimistic, whereas "Job Losses" clearly is a pessimist's term. An appropriately brazen title offers a good way to overstate economic health or industrial illness.

If the picture is bleaker or brighter than suits your politics, try a rate of change rather than a mere rate. After all, minor downturns often interrupt a run of good years, and depressions do not last forever. If unemployment is high now but a bit lower than a year, six months, or a month ago, the optimist in power would want a map showing a significant number of areas with declining unemployment. Conversely, the pessimist who is out of power will want a map depicting conditions at least as bad as before the current scoundrels took over. A time interval that begins when proportionately fewer people were out of work will make the opposition party's point, especially if unemployment has become worse in large, visually prominent, mostly rural regions.

A useful index for the optimist is one with relatively low values, such as the unemployment rate, if conditions have improved, or an index with comparatively high values, such as employment level, if conditions are worse. Thus a drop of one percentage point from a base of 4 percent unemployment yields an impressive 25 percent improvement! Yet a substantial increase in the unemployment rate from 4 to 6 percent can be viewed more optimistically as a drop in labor force participation from 96 to 94 percent—a mere 2 percent drop in employment.

Point symbols and counts, rather than rates, can be useful too. If the economy has been improving in all regions, the current government might want a map with graduated circles or bars showing actual counts beneath the title "Employment Gains." If the country is in a widespread recession, the opposition would use similar point symbols with the title "New Job Losses."

The cartographic propagandist is also sensitive to spatial patterns. Favorable symbols should be large and prominent, and unfa-

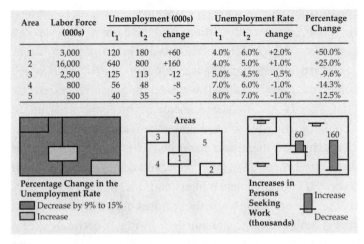

Area	Labor Force (000s)	Unemployment (000s)			Unemployment Rate			Percentage Change
		t_1	t_2	change	t_1	t_2	change	
1	3,000	120	180	+60	4.0%	6.0%	+2.0%	+50.0%
2	16,000	640	800	+160	4.0%	5.0%	+1.0%	+25.0%
3	2,500	125	113	-12	5.0%	4.5%	-0.5%	-9.6%
4	800	56	48	-8	7.0%	6.0%	-1.0%	-14.3%
5	500	40	35	-5	8.0%	7.0%	-1.0%	-12.5%

Figure 11.19. Unemployment data (top) for a hypothetical region (bottom center) yield different maps, supporting an optimistic view (bottom left) and a pessimistic view (bottom right) of recent temporal trends.

vorable ones small and indistinct. Thus the optimist might present the unemployment data in figure 11.19 with the map at the lower left, to focus attention on improved conditions in larger areas, whereas the pessimist would prefer the map at the lower right, to emphasize the much greater number of unemployed persons in more urban areas. Note as well how the titles and keys in these examples reinforce cartographic manipulation.

Labor economists, who commonly adjust unemployment data for seasonal effects, discourage some manipulation of time intervals. After all, more people are seeking work in early summer, when many high-school and college graduates enter the labor force for the first time. And more people find at least temporary work in November and December, the peak holiday shopping season. Local seasonal effects, such as tourism and the temporary hiring of field and cannery workers in agricultural areas, also require seasonal adjustment.

Mortality, fertility, and other phenomena that do not affect all segments of the population equally also require adjustment. Figure 11.20, a comparison of the age-adjusted death rate with the crude

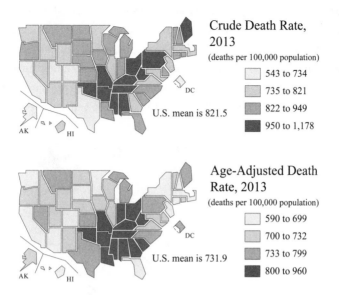

Figure 11.20. Maps of the crude death rate (top) and the age-adjusted death rate (bottom) can present markedly different geographic patterns of mortality.

death rate, illustrates the wisdom of mapping demographically adjusted rates. The map at the top is a simple rate, which does not consider, for example, Maine's relatively older population. When the rates portrayed in the upper map are adjusted for age differences, Indiana and several southeastern states emerge as high-rate areas whereas those in the Northeast slip to a lower category. Age-adjustment allows the map at the bottom to reveal the effects of relatively good health care and a higher socioeconomic status in the New England, Middle Atlantic, and North Central states, in contrast to greater poverty and less accessible health care in the South.

Be skeptical of maps based on numbers. Because a single variable can yield many different maps, don't be silenced by the argument that more than one map would cause needless confusion. You might ask to see several maps, or be given the opportunity to experiment with categories and symbolization online or using software.

Be wary of not only the known cartographic manipulator but also the careless map author unaware of the effects of aggregation and classification. Also question the definitions, measurements, short-cuts, and motives of a government agency, research institute, or polling firm that generously provides its data—even the most con-scientious mapping effort is undermined by flawed data.

Image Maps: Picture That

Image maps are pictures, much like digital photos. They originated with airborne photography, whereby a camera carried aloft captured an image of the landscape on a light-sensitive sheet of glass or film. The addition of lines or labels makes an air photo a full-fledged map, and electronic computing adds further refinement once a scanner converts the image to a grid of *pixels*, or picture elements, each with a numerical code describing a small area on the ground. Digital image maps can be displayed on a computer monitor or other electronic device, projected onto a screen, printed on paper, enhanced electronically to heighten contrast or highlight selected features, and transformed geometrically to remove relief displacement (see chap. 3). The genre includes facsimiles of historic maps scanned by collections eager to share online and digital pictures from orbiting instrument packages and weather-radar networks.

This chapter looks briefly at the structure, content, and uses of image maps: how they differ from other maps, how they're developed, and how they can reveal as well as distort what is commonly accepted as truth. Improperly annotated, they might even start wars.

Grids, Sensors, and Platforms

Image maps are developed, stored, processed, and displayed as *raster* data, a term electrical engineers derived from the Latin word for *rake* to describe the parallel lines formed inside a cathode ray tube. Raster data typically are organized as a grid of numbers or pixels arranged in rows and columns, in contrast to vector data, whereby map features are represented by lists of points described by rectangular (X, Y) coordinates. Raster data are ideal for describing all parts of a region, with one or more numbers for each pixel, whereas vector data are better for the precise portrayal of individual features like roads and boundaries.

Plate 12 underscores the fundamental difference between vector and raster representations. Both images describe the same area, in Bath, Maine, along the Kennebec River. The left-hand excerpt is a conventional topographic line map, so called because its symbols are largely elevation contours and lines representing roads, streams, shorelines, and political boundaries as well as the perimeters of built-up areas (in light red) and patches of woodland (in green). The right-hand excerpt is an orthophotomap, corrected to remove relief displacement. It offers a more detailed, less interpreted picture of the landscape and a more informative picture for anyone interested in the Bath Iron Works, a major shipbuilding firm along the river's west bank, just south of the bridge. Look carefully and you can see several naval vessels under construction or in for repair, and elsewhere on the image map you can see field boundaries and other discontinuities in land cover missing from its topographic counterpart, which is more adept at pointing out City Hall, schools, churches, and other landmark structures. As this example confirms, line maps and image maps are not only different but complementary.

Like any map, an image map has the basic cartographic elements of scale, projection, and symbolization, which reflect its overall level of detail, the degree of stretching or compression required to flatten the planet's curved surface onto a sheet of paper or a display screen, and the colors or graytones used to describe spatial variation. No less important is the image map's resolution, typically expressed as the length on the ground of one side of a square represented by a pixel. Although smaller is generally better, an enormous grid of tiny ground-resolution cells can put needlessly heavy demands on electronic memory and computer processing.

What's appropriate depends on the purpose of the analysis as well as the phenomenon portrayed: ground cells that are 10 centimeters on a side might be required for military intelligence, whereas ground cells that are 10 meters on a side could be fully adequate for an evaluation of forest resources across a multistate region. (Metric measurement has been standard in satellite remote sensing from the outset.) An intelligence officer needs to distinguish between cars, trucks, and tanks, while an environmental scientist might be interested only in large stands of trees. Because interactive display systems let the viewer zoom in or out, the scale of the image on the screen is less relevant than its ground resolution, which limits how small an object a sensor can detect.

What's shown on the screen or paper is the conversion to light, ink, or toner of one or more measurements recorded for each ground-resolution cell. When the image map is a conventional black-and-white aerial photo, the pixel records the amount of light reflected from the ground in the visible part of the electrometric spectrum, as described in the left half of figure 5.1. Because the shorter wavelengths, in the blue part of the visible band, are too readily scattered by the atmosphere, they are commonly filtered out to yield a less hazy image—although panchromatic imagery in principle represents all wavelengths of visible light, sharpness requires a bias toward green, yellow, orange, and red.

Black-and-white image maps are a diverse lot: some are captured and reproduced using only photosensitive emulsions, others are converted to digital data by a tabletop scanner, and some are "born digital" using a digital camera or an instrument known as a radiometer, which can record differences in reflected radiation outside the visible band (wavelengths 0.4 – 0.7 µm). Particularly important is an emulsion's or radiometer's sensitivity to energy in the reflected infrared part of the spectrum (roughly 0.7 – 2.5 µm), just beyond what the human eye sees as red light. Near-infrared sensitivity is important because some land covers, notably healthy vegetation, reflect much of the near-infrared radiation they receive from the sun, whereas other types, including the camouflage netting used during World War II to hide artillery and other equipment from reconnaissance aircraft, have a markedly lower infrared reflectivity. Devised by military scientists, camouflage-detection film yields photo prints showing real vegetation in light tones and fake vegetation as comparatively dark—a dead giveaway where the texture suggests trees and shrubs.

Camouflage-detection film and electronic photography eventually led to multispectral scanners, which are adept at distinguishing between vegetation, bare soil, snow, and clear water, with the distinctive "spectral signatures" shown in figure 12.1. Because multispectral scanners record reflected energy for comparatively narrow parts of the spectrum, healthy vegetation, which looks green when only visible light is recorded, registers a dominant reflectance in the near-infrared band, or channel, with recorded wavelengths typically between 0.7 and 1.1 μm. Although black-and-white prints can be made using only the measurements in the sensor's near-infrared channel, scientists devised a type of image known as color infrared (CIR for short) but also dubbed "false color" because the three primary colors on a video monitor (red, green, and blue) are assigned respectively to the near-infrared, red, and green portions of the spectrum. This spectral shift makes areas with healthy green vegetation appear bright red because their near-infrared reflectance dominates whatever red and green light that trees and crops reflect upward. By contrast, urban and suburban areas, which mix radiation reflected from paved surfaces and roofs as well as from trees and green grass, typically appear gray or blue-gray; bodies of water, which reflect comparatively little green, red, and infrared radiation, look nearly black; and snow and any puffy clouds that drift into the scene look white because they reflect strongly in all channels.

These contrasts are dramatically apparent in plate 13, a satellite image depicting the aftermath of the tornado that rampaged through Tuscaloosa and Birmingham, Alabama, on April 27, 2011, ripping out trees and buildings. The beige stripe running from low-

Figure 12.1. Spectral signatures for bare soil, vegetation, and water in the visible and near-infrared portions of the electromagnetic spectrum.

er left to upper right marks the storm's path, reported by ground observers to be 1.5 miles wide in places. The white rectangular areas are most likely metal roofs that reflect strongly in all of the sensor's channels.

These attempts to measure the relative amount of light radiated from below in various parts of the electromagnetic spectrum spawned a complex technology known as remote sensing. Even before the first Landsat was launched in 1972, remote-sensing technologists were experimenting with the wavelengths that bracket a sensor's various channels in order to optimize detection of specific land covers, such as rough pasture, deciduous trees, conifers, and various grasses and field crops, and to assess soil moisture conditions, geologic structure, and the vigor of trees and field crops. Improvements included additional channels, which can be ignored, merged, or otherwise interrelated, and image-processing software for heightening contrast, sharpening boundaries, comparing different channels, and classifying land cover by comparing pixels with "ground truth" values known to represent specific categories.

Remote sensing involves both an imaging system and an overhead platform like an airplane, drone, or satellite. In general, airplanes capture imagery from lower altitudes by flying back and forth across a project area in overlapping strips, while satellites operating from a much greater altitude provide repeat coverage over a vast area with a generally larger, less-precise ground resolution. The ASTER (Advanced Spaceborne Thermal Emission and Reflection Radiometer) satellite, which captured the image in plate 13, scans along a moving ground swath 60 kilometers (37 miles) wide designed to follow the sun by crossing the equator at 10:30 am local time, always on a north-to-south pass inclined 98.3° from the equator, which accounts for the left and right edges not perpendicular to the illustration's base. An altitude of 705 kilometers and an orbit period of 98.88 minutes is sufficient to balance forces that would bring down a slower satellite or fling a faster one into outer space. In addition to scanners looking directly downward as well as backward, the satellite also has a pivoting telescope able to point sideways, to image selected areas more frequently than the ground swath's sixteen-day repeat interval would allow. Multiple sensors capture imagery with ground resolutions of 15, 30, and 90 meters; the larger pixels are for a thermal scanner, useful for monitoring drought, air pollution, and weather conditions.

Large ground-resolution cells are typical of geostationary satellites that look downward from 35,786 kilometers (22,236 miles) above a constant position on the equator and make one complete revolution around the Earth's axis every twenty-four hours. Well suited to take snapshots of cloud cover and atmospheric moisture every fifteen minutes across almost a third of the planet, their instrument packages track hurricanes, detect emergent forest fires, and assess the impacts of climate change. At much lower altitudes satellites in a near-polar orbit collect timely information about ice sheets, the oceans, and the upper atmosphere. Image maps of physical phenomena such as soil moisture, radiation heat loss, and reflectivity of light in different parts of the electromagnetic spectrum made overhead imaging one of the twentieth century's two or three truly noteworthy cartographic revolutions.

Figure 12.2. Before (left) and after (right) views of the Bath, Maine, orthophotomap (plate 12, right) selectively altered with Photoshop.

Blurring and Mislabeling

Image maps rarely display raw measurements. For example, satellite imagery captured for a skewed array of overlapping, roughly circular "ground spots" is typically resampled to a uniform grid of square pixels, and imaging specialists commonly tweak the intensities of a sensor's channels to highlight specific types of land cover. Colored image maps that are based on orbiting or airborne sensors with more than three channels are very much authored views of the landscape and occasionally resemble abstract paintings.

Possibilities for benign or malignant deception abound because imaging specialists have digital tools similar to the software with which twenty-first-century portrait photographers correct crooked noses, fix yellowed or missing teeth, and remove wrinkles, acne scars, and aging spots. National defense might justify blurring or erasing military targets like rooftop surface-to-air missiles, and privacy concerns might require the blurring of nude sunbathers or outdoor fornicators on very high-resolution online image maps. Because blurring is a telltale sign that attracts further scrutiny, careful redacting is needed to preserve a reassuring sense of authenticity. Although viewers unaccustomed to close analysis are easily fooled, a forensic scientist skeptical of photographic evidence could readily uncover the shenanigans in the right part of figure

12.2, in which I altered the orthophoto of Bath, Maine, by erasing the bridge and city hall, moving a ship northward to what had been the west end of the bridge, and converting a parking lot to wooded land. Unless a broad conspiracy is afoot, skeptical viewers might want to compare multiple images acquired at different times.

Because features of interest are often not obvious to the untrained eye, image analysts add authoritative annotations, which viewers tend to believe unless they know otherwise or are cautiously skeptical. Where doubts arise about the source's motives, a strong endorsement from a trusted authority can stifle suspicion, as on February 5, 2003, when the US Secretary of State, retired general Colin Powell, testified before the United Nations Security Council that Iraq's president Saddam Hussein had been manufacturing and stockpiling "weapons of mass destruction" (WMDs) and had hid them before the arrival of UN inspectors. Powell illustrated his presentation with high-resolution satellite imagery. One of his slides (plate 14) showed "a weapons munitions facility… at a place called Taji," where multiple bunkers contained "chemical weapon shells." The close-up in the left-hand panel shows "sure signs that the bunkers are storing chemical munitions." The small facility labeled "Security" is "a signature item for this kind of bunker." Inside are "special guards and special equipment to monitor any leakage" that might occur. Another "signature item" is "the decontamination vehicle in case something goes wrong." On the right, an image captured six weeks later, when the UN inspection team was arriving, shows "sanitized bunkers" with their telltale decontamination vehicles removed. "The bunkers are clean when the inspectors get there [and] they found nothing."

Powell believed the annotated images he had been given. Before presenting his evidence, he conceded, "The photos I am about to show you are sometimes hard for the average person to interpret, hard for me." Powell trusted the "experts with years and years of experience," who had been "poring for hours over light tables." His presentation was convincing, but the subsequent invasion, backed by the UN and the US Congress, found no WMDs, which suggests that the interpretation was seriously flawed, perhaps because American and British officials were focused on unseating the Iraqi dictator. The Iraq War, sometimes called the Second Persian Gulf War, lasted until 2011, far longer than expected. It toppled Saddam Hussein but destabilized the Middle East, cost tens of thousands of lives and billions of dollars, triggered a massive outpouring of refugees, and demonstrated that image maps are too easily decorated with labels that confirm what leaders want to believe.

Labels like those in Powell's PowerPoint slides reflect interpretation, not knowledge, and flawed interpretation can radically alter the meaning of an otherwise mundane image. Plate 15, which paraphrases a droll reinterpretation suggested by artist Daniel Mooney, is a succinct warning that image maps can be mislabeled by biased analysts. Motives are important: we can more reliably believe that the "City Hall" in the topographic-map excerpt is actually the home of municipal offices because we trust the mapmaker—but too much is at stake if lax editing at the Geological Survey were to empower Michigan-football fans (e.g., fig. 4.4) or similar mischief-makers. When stakes are high, be careful what you accept at face value.

Prohibitive Cartography: Maps That Say "No!"

Maps have a long history of saying, "Stay out!" or "Don't go there!" or "Don't do that!" Classic examples are border lines that defend against invaders and property lines that ward off trespassers—although lines on maps are only graphic symbols, the implied threat of military or legal action gives them meaning. People have learned to respect boundary and property maps, and as governments became ever more organized and controlling in the twentieth century, restrictive maps became ever more numerous and diverse. With this heightened credibility, they are ripe for deception.

Prohibitive cartography operates at multiple scales. Small-scale, wide-scope prohibitive maps focus on international borders, which began appearing offshore well before 1982, when the United Nations Convention on the Law of the Sea authorized nations to claim jurisdiction over fishing, seabed mining, and marine conservation within Exclusive Economic Zones (EEZs) extending 200 nautical miles (230 statute miles or 370 kilometers) beyond their respective shorelines. Maps dividing up a significant chunk of the high seas also demonstrated the formidable geopolitical leverage of island territories, with dramatic consequences like the offshore boundary between Japan and the United States, which had acquired the Northern Mariana Islands as a "trust territory" after World War II (fig. 13.1). The US Department of State has a small but essential cartographic unit responsible for the concise delineation of borders and territorial claims. Cartographic evidence is used to announce, argue, and settle claims, now recorded as lists of geographic coordinates (latitudes and longitudes) marking the "turning points" at

Figure 13.1. Exclusive Economic Zones carved out of the western Pacific Ocean by the United States (dark gray) and other nations (white) underscore the importance of the comparatively tiny islands to which much of this maritime territory is anchored.

which straight-line boundary segments meet. Closer to shore, the territorial sea, once only 3 miles wide along most shorelines, now extends outward 12 nautical miles, and most countries also claim a "contiguous zone" reaching outward another 12 miles.

At the more detailed, smaller-scope end of the spectrum of map scales are property maps and zoning maps, both of which are diluted by exceptions not revealed on parcel surveys. Although property maps typically describe boundary lines, street lines, and the locations of driveways and structures, they sometimes omit rights-of-way that afford legal access to a neighboring property, easements for overhead wires or buried pipes and cables, and encroachments such as an over-the-line porch installed years ago by a neighbor who misread (or ignored) a deed or survey—in some states, statutes on "adverse possession" give that neighbor a permanent right to part of your land if the encroachment was visible and you did not object. Another restriction on property rights is the regional database that inventories buried infrastructure: anyone planning a fence, deck, in-ground pool, or regraded lawn must notify the local "one-call" center, which will send out a surveyor to mark the locations of any water, sewer, gas, telecommunications, and electric lines. If the intended excavation would interfere with underground utilities, you must adjust your plans before moving forward.

Zoning maps (fig. 7.1 and plate 5) impose other restrictions on property owners, who usually appreciate the freedom from noise and traffic afforded by a low-density, highly residential R-1 zone, which bans commercial and industrial activity as well as apartment buildings and multifamily residences. Land-use zoning typically imposes a host of restrictions, such as the maximum height of structures, mandated setbacks from a lot's front, side, and back boundaries, the types and number of animals a resident may keep, and the number and kinds of vehicles that can be parked in the driveway. Although zoning maps are usually reassuring—few people want to live next to a jalopy hoarder or personal petting zoo—a neighboring parcel's zoning category might entitle its owner to run a specific kind of small business or boarding house. For example, the statutes might allow a medical or dental office in the practitioner's residence or sanction a group home for developmentally or physically disabled residents and their caregivers. Zoning laws typically allow a relaxation of setback requirements or other exceptions after a proposed "variance" has been advertised to nearby residents, discussed at a public hearing, and approved by the Zoning Board of Appeals. Zoning maps are significant not only for what they restrict but also for any exceptions the law might allow.

Restrictive maps are essential to historic districts, which create a sacred space of sort for preservationists, including residents who want to protect their investment in buildings and streetscapes from commercial tackiness and mundane modifications. Accepting the umbrella of a historic district narrows one's choice of replacement windows, doors, and siding and precludes structural modifications usually allowed outside the district. If the rules require white clapboard siding and black wooden shutters, anything else could trigger a messy lawsuit. Outside the boundary, you're OK. Inside, you're stuck, which is fine if that's your preference.

Another kind of prohibitive map delineates agricultural districts that protect farmers from a new neighbor's complaints about obnoxious sights, sounds, and smells. Agricultural districts delineated on maps not only guarantee a landowner's "right to farm" but also defend against higher property taxes based on the property's market value. Owners might need to give up development rights, thereby sacrificing substantial profit decades ahead if an entrepreneur wants to buy the land for a shopping center or a housing development.

Restrictive cartography is embedded in most environmental regulations intended to protect habitats or home buyers (see plate 9). Examples include maps of protected wetlands, maps showing where licensed hunters can "harvest" deer during a designated "season," and flood-zone maps indicating where structures are prohibited and who must buy flood insurance. All three types of map are inherently fluid, largely but not solely because of Nature's whims. Wetlands maps begin with a formal delineation, following federal or state guidelines, with state regulations relevant only if a wetland is not part of the "waters of the United States." A delineation can expire and might need to be redone if carried out during a drought. Different states define wetlands differently, and sometime a developer can create a substitute wetland elsewhere if a protected site is essential to a larger development plan. State agencies, which administer wildlife management plans, can adjust the length and geographic scope of deer season to the herd's size and the need for culling.

Flood zones are particularly problematic. Based on a modeling process that relies on streamflow data collected at a small number of gauging stations and on elevation data that are often imprecise, flood maps are readily challenged as inaccurate and unreliable by people who don't want to buy flood insurance, which lenders require. Although complex and potentially confusing map keys, as in figure 13.2, try to explain that "hundred-year flood" refers only

LEGEND

 SPECIAL FLOOD HAZARD AREAS (SFHAs) SUBJECT TO INUNDATION BY THE 1% ANNUAL CHANCE FLOOD

The 1% annual chance flood (100-year flood), also known as the base flood, is the flood that has a 1% chance of being equaled or exceeded in any given year. The Special Flood Hazard Area is the area subject to flooding by the 1% annual chance flood. Areas of Special Flood Hazard include Zones A, AE, AH, AO, AR, A99, V, and VE. The Base Flood Elevation is the water-surface elevation of the 1% annual chance flood.

ZONE A	No Base Flood Elevations determined.
ZONE AE	Base Flood Elevations determined.
ZONE AH	Flood depths of 1 to 3 feet (usually areas of ponding); Base Flood Elevations determined.
ZONE AO	Flood depths of 1 to 3 feet (usually sheet flow on sloping terrain); average depths determined. For areas of alluvial fan flooding, velocities also determined.
ZONE AR	Special Flood Hazard Areas formerly protected from the 1% annual chance flood by a flood control system that was subsequently decertified. Zone AR indicates that the former flood control system is being restored to provide protection from the 1% annual chance or greater flood.
ZONE A99	Area to be protected from 1% annual chance flood by a Federal flood protection system under construction; no Base Flood Elevations determined.
ZONE V	Coastal flood zone with velocity hazard (wave action); no Base Flood Elevations determined.
ZONE VE	Coastal flood zone with velocity hazard (wave action); Base Flood Elevations determined.

FLOODWAY AREAS IN ZONE AE

The floodway is the channel of a stream plus any adjacent floodplain areas that must be kept free of encroachment so that the 1% annual chance flood can be carried without substantial increases in flood heights.

OTHER FLOOD AREAS

ZONE X	Areas of 0.2% annual chance flood; areas of 1% annual chance flood with average depths of less than 1 foot or with drainage areas less than 1 square mile; and areas protected by levees from 1% annual chance flood.

OTHER AREAS

ZONE X	Areas determined to be outside the 0.2% annual chance floodplain.
ZONE D	Areas in which flood hazards are undetermined, but possible.

Figure 13.2. Upper half of a typical explanation panel illustrates the complexity of flood-insurance-rate maps, used to inhibit development and determine insurance rates on floodplains.

to a flood level with an annual probability of 1 percent, many people don't understand that the occurrence of a hundred-year flood last year does not guarantee ninety-nine risk-free years. Moreover, if extensive new land development has occurred upstream, flood maps can significantly underestimate the effects of rapid runoff from roofs, roads, and other impervious surfaces.

A particularly invidious threat is accelerated sea-level rise associated with climate change, which promises more frequent severe coastal storms as well as the eventual inundation of low-lying areas worldwide. Because rising seas are a slow phenomenon for which *whether* is easier to forecast than *when*, climate scientists are wary of making detailed predictions. Unable to ignore the hazard's high consequences, they responded cautiously, with coastal-elevation maps depicting areas likely to be underwater should, for instance, the seas rise 1.5 meters (nearly 5 feet) by the end of this, or the next, century. Maps like these inspire diverse reactions, with self-styled skeptics condemning them as meaningless hoopla, optimists hoping that increased awareness of the threat will inspire global cooperation in curtailing greenhouse gases, and pessimists calling for mandatory restrictions as well as a massive retreat from the current shoreline. Coping with uncertainty has become a serious concern for mapmakers.

Perhaps the quintessential prohibitive map is the aeronautical chart, which both creates and regulates airspace. Restricted areas around weapons-testing sites warn pilots of hazards from below; prohibited airspace defends the White House, nuclear plants, and Disneyland against tragic accidents that could kill thousands or seriously damage the national psyche; and maps showing designated flight paths help air-traffic controllers prevent mid-air collisions. Airspace regulations now include maps of temporary flight restrictions (TFRs), which pop up online as needed to safeguard the movements of top officials, warn aircraft away from hazards like wildland fire and fretful volcanoes, and provide security for large outdoor concerts, environmental protests, and other large gatherings.

Cartographically announced restrictions also include "no-fly" zones over wide areas like Bosnia-Herzegovina (1993–1995) and Iraq (1991–2003). Military no-fly zones typically link maps of war zones or national territory to complicated rules of engagement, which can make these restrictions little more than a hollow threat—the geopolitical counterpart of a spanking threated by a parental disciplinarian who prefers persuasion to abuse. More local, less-combative no-fly zones emerged in response to commercial development of unmanned aerial systems (UASs) in the 2010s. Prohibitive maps that don't trigger an assured, immediate response can still be useful in controlling bad behavior. It's the nature of the state to discipline its citizens, and mapping is part of the process.

Restrictive cartography could have a far darker future under an authoritarian regime able to declare and enforce "no-go" areas with the aid of GPS tracking systems and "geofencing" enforced with an electronic pinch—software would monitor compliance by comparing the subject's continuously updated location with a perimeter stored in a GIS. Because geofencing magnifies the map's ability to manipulate behavior, informed skepticism is more prudent than the outright rejection of social scientists who warn of the "surveillance state."

Fast Maps:
Animated, Interactive, or Mobile

Around the turn of the twenty-first century, maps moved online. The change has been massive, much like the digital revolutions that swept the news business, entertainment, and political discourse, all of which at times seem curiously conflated. Scholars and pundits assert that maps are now ubiquitous, an obvious exaggeration but not far from the truth. More significant is that maps have become remarkably fast in both their creation and dissemination, and so strikingly that twenty-first-century electronic maps collectively merit the label *fast maps*, mostly because no other general term adequately accounts for both their velocity and their diversity. These maps can deceive in the traditional ways that have been discussed throughout this book, but they also raise unique issues that never arose with static or paper maps.

After concise looks at the diversity of fast maps and the technology that forced mapping firms to change their business model, their mission, and the names and terms they use, this chapter summarizes the advantages and challenges posed by this new way of making, delivering, and using maps.

Diversely Interactive

Fast maps are a diverse lot. Many, perhaps most, are in some way interactive insofar as the user can alter at will their content, symbols, and level of detail. Some interactive maps provide travel directions; others are a significant part of the graphics software used by data analysts to explore geographic influences on various phenomena. Maps also provide an interactive index to vast troves of geographic data compiled for applications as diverse as locating retail outlets, predicting election results, understanding climate change, or planning vacations and business trips—tasks for which the map is an appropriate "human-machine interface." Cartographic interactivity can also be passive, as when a satellite navigation system or mobile device focuses relentlessly on the viewer's progress along a highway, across a field, or through a building while momentarily relevant graphic symbols parade past the you-are-here spot at the center of the screen. Indeed, a key challenge of interactive cartography is the constraint posed by small screens; this is addressed mostly by software that allows zooming in, zooming out, and panning around.

Not all dynamic maps are interactive. Animations crafted to describe a process or historic event might neither require nor tolerate viewer input, and, if allowed, viewer interaction might be limited to the selection of the region shown or of the map's content or timeframe, as when animated weather maps describe the past or predicted movement of storm systems. Other fast maps are stodgily static in content and appearance but indisputably fast when timely or provocative content, or a quirky design, quickly wins them a vast audience. Static web maps that "go viral" on social media can advance a point of view, embarrass a political opponent, or galvanize public disgust by exposing radical disparities in real property assessments, dramatizing a public figure's romantic affairs or suspicious real-estate deals, or revealing an alarming pattern of rapes or shootings.

Cartographic outing gets people's attention. The vast amount of public data on crime, property ownership, and gun registrations presents an intriguing opportunity for exposing alleged criminals or highlighting the geography of handgun registrations—perhaps with unintended consequences. After a newspaper in a New York City suburb used a FOIA (Freedom of Information Act) request to obtain an address listing of pistol permits in two counties, it posted a zoomable map showing who in one's neighborhood might own a handgun (plate 16). Outraged gun owners complained that the data not only ignored assault rifles and shotguns but made permit holders, who might not have actually purchased a firearm, vulnerable to intrepid burglars, no doubt the worst kind. Despite applause from gun-control advocates, the newspaper took down the map and hired armed guards to protect its staff and property.

Crowdsourcing, Map Tiles, and the Web Mercator Projection

Fast maps reflect the rise of a host of technological innovations, most notably digital computing, increasingly rapid microprocessors, high-resolution display screens, cheap and commodious electronic memory, satellite positioning, wireless telecommunications, and the internet and its versatile subset, the World Wide Web. Web cartography not only responds to requests for maps and directions but also supports volunteered geographic information, also known as crowdsourcing. Examples include OpenStreetMap, a service enhanced by users who report new streets and trails, and community-based motor-vehicle navigation systems that use otherwise unexplainable localized reductions in speed to identify and report traffic congestion. Some in-vehicle navigation systems not only respond to congestion by recommending alternative routes but also allow drivers to divulge the emergence and disappearance of speed traps and police checkpoints—an ethical quandary perhaps, but these disclosures make many motorists slow down. Direct intervention is also possible if GPS tracking and enhanced highway databases are allowed to proactively slow down vehicles flagrantly exceeding the posted speed limit or halt a truck moving ominously toward a low bridge or tunnel with inadequate vertical clearance. Proactive intervention is essential for driverless cars, which can increase the capacity of existing roads as well as take the fun out of driving.

Web cartography depends on a storage and retrieval strategy known as *tiling*, which divides an electronic world map into a large number of small rectangles known as tiles. Figure 14.1 describes

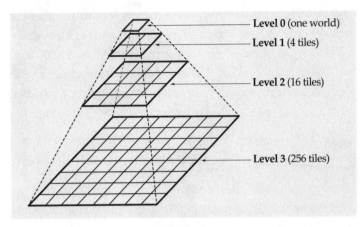

Level 0 (one world)

Level 1 (4 tiles)

Level 2 (16 tiles)

Level 3 (256 tiles)

Figure 14.1. Zoomable web maps divide tiles into smaller pieces as the viewer zooms in for a closer, more detailed look.

this approach as a succession of zoom levels that start with a whole-world view at the top of a pyramid for which each closer, more detailed step involves markedly more tiles, each covering less territory in greater detail. To support the rapid zooming of web maps, sets of tiles must be generalized and stored for as many as twenty-two specific levels of detail. A user requesting a map for a particular address receives a display several steps up from the most detailed level. Clicking the plus sign to zoom in or the minus sign to zoom out typically reveals more or fewer types of features. A jump in the map's labels is also apparent, as when street and business names appear or disappear when zooming in or out, as exemplified by the three successive levels in plate 17. Boundaries between tiles are never visible, and the number of tiles does not increase by a factor of 4, as figure 14.1 might imply.

For each zoom level, linear features such as roads and boundaries have been selected and displaced as needed to avoid clutter but made to meet seamlessly along boundaries between adjoining tiles. When the user drags the map across the screen, the browser retains some tiles from the previous display and adds new ones along the old map's trailing edges. Because efficient micropro-

cessors make panning appear seamless, as if the map is slipping
across the screen, tiled web maps are sometimes called *slippy maps*.

A subtle consequence of web-map tiling is an invigorated use of the Mercator projection, denounced roundly for its infamous distortion of area on small-scale world maps but an indispensable tool for marine navigation and field artillery in the pre-satellite era (see chapter 8). The Mercator map is well suited to tiling because it does not distort angles. This mathematical property, called *conformality*, allows only a minimal, largely unnoticeable distortion of local shapes and distances on individual tiles at the more detailed zoom levels. The Mercator map also provides a rectangular grid of meridians and parallels, which accommodates the rectangular screens of laptop computers and mobile devices as well as the massive electronic storehouse of tiles, served up rapidly when a viewer pans across the map or requests more detail. A single projection assures an acceptably smooth transition from one zoom level to the next.

Cartographers attuned to the mathematical details of map projection acknowledge that the Mercator projection used for web mapping is not quite the same Mercator projection used for large-scale topographic maps, which are cast on an ellipsoid rather than on a sphere. Ellipsoids, which provide a more accurate geometric framework for military and scientific applications, are spheres flattened strategically at the poles to more closely approximate Earth's actual shape. Extreme accuracy demands different ellipsoids for different continents, but this precision would be lost on web cartography, which gets by quite well by assuming a less computationally demanding spherical planet. Purists call this approximation the "Web Mercator projection" and note the use of different mathematical shortcuts—which means nothing to most users.

More problematic is the fully zoomed-out whole-world map, as in figure 14.2, on which Greenland looks inappropriately similar in size to South America. Coping strategies available to web-map services include naive acceptance of the traditional Mercator worldview (conveniently truncated to avoid a pointlessly humongous

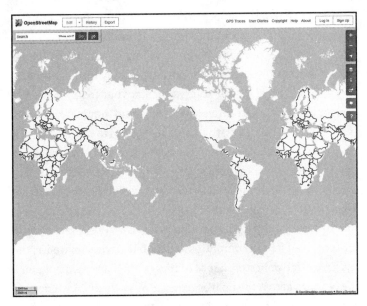

Figure 14.2. OpenStreetMap zooms out to a continuously pannable, uninterrupted Mercator map that omits extreme northern and southern latitudes.

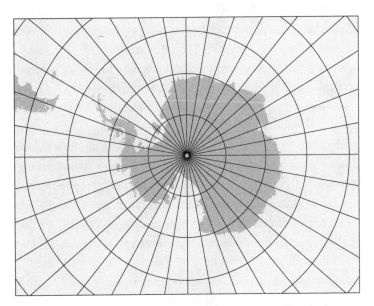

Figure 14.3. Antarctica would be better served by a polar stereographic projection, which is conformal.

Antarctica), restricting the degree to which viewers can zoom out, and jumping to a more appropriate *nonrectangular* geometric framework for a whole-world map enhanced with environmental data that include polar regions. Even so, any cylindrical projection—not just the Mercator—is wholly inadequate when a viewer wants to zoom in on Antarctica, which is better served by a polar conformal projection, as in figure 14.3.

Benefits and Caveats

Whether online or not, fast maps offer many benefits, including improved methods for visualizing uncertainty. Possibilities abound when a viewer can explore the fuzziness of the data by changing a map's time frame or definitions, or question the reliability of a choropleth map by experimenting with cut points or number of categories. Meteorologists in particular benefit from a diverse choice of atmospheric models, which can be altered as well as compared cartographically in order to assess the reliability of a forecast. In many cases, a prediction is reliable only when those affected understand the underlying uncertainty.

The benefits of *fast* are readily apparent when a web map is linked to a search engine designed for instant response. Type in an address or the name of a business and a map appears in a few seconds, and it's free. Or is it? A web map service (WMS) is a business supported by venture capitalists confident that advertising revenue will not only cover operating costs but earn them a decent return on their investment. An efficient, well-promoted WMS attracts millions of viewers, whose attention is sold to advertisers, whose messages appear on the map's margins, and sometimes on the interior, as when brand-name coffee bars and hotel chains pay to put their logos on a map rather than be represented by a generic icon like a coffee cup or bed. Putting a business's name on the map is free, but a familiar icon increases the likelihood a viewer will click to find the phone number, determine hours of operation, or make it a destination for the GPS. A firm can add also an interactive street map to its website—perhaps for free, unless the map is heavily used.

Although a single WMS might dominate the market, competitors that own a popular search engine or otherwise attract significant internet traffic can survive nicely. To increase market share, firms add enhanced services like street-view images or try to make their maps more informative or visually attractive. Innovations can become controversial, as when Google Maps began using a pale-orange tint to highlight areas relatively rich in things to do: neighborhoods toward which casual viewers exploring a city for the first time might want to zoom in for a closer look. For example, the tinted area in the top-left panel in plate 17 shows only one business, while the bottom panel, two clicks closer in, reveals many more potential destinations. What's problematic is the suggestion, not rigorously confirmed, that Google might have ignored equally vibrant clusters of ethnic or minority-owned businesses in other parts of the city. Could an unintended bias be embedded in the algorithm that defines these areas of interest? Perhaps, but far more troubling is the involuntary exploitation of whatever knowledge of motivations and personality might be inferred from a viewer's interaction with web maps.

Are fast maps luring the public away from paper maps? Probably, but no more so than television, streaming video, and motion pictures are replacing books and magazines. Bookstores remain lively places, and every good bookstore has a rich offering of attractive, cleverly designed folded maps and atlases that can be savored in ways impossible with a computer's or mobile device's small screen. If slow maps eventually become a niche market, they will retain an enthusiastic following.

Epilogue

The preceding chapters have explored the wide variety of ways in which maps can lie: why maps usually must tell some white lies, how maps can be exploited to tell manipulative lies, and why maps often distort the truth when a well-intentioned map author fails to understand cartographic generalization and graphic principles. The wise map user is thus a skeptic, ever wary of confusing or misleading distortions conceived by ignorant or diabolical map authors.

Let me conclude with a cautionary note about the increased likelihood of cartographic distortion when a map must play the dual role of both informing and impressing its audience. Savvy map viewers must recognize that not all maps are created solely to inform the viewer about location or geographic relationships. As visual stimuli, maps can look pretty, intriguing, or important. As graphic fashion statements, maps not only decorate but send subtle or subliminal messages about their authors, sponsors, or publishers. Some advertising maps, for instance, announce that a power company or chain restaurant is concerned about the city or region, whereas free street guides attest to the helpfulness of a real-estate firm or a bank. A flashy map, in color with an unconventional projection, touts its author's sense of innovation, and cartographic window dressing in a doctoral dissertation or academic journal suggests the work is scholarly or scientific. An ornate print of an eighteenth-century map of Sweden not only decorates a living-room wall but proclaims the household's pride in its Scandinavian heritage. A world map behind a television newscaster reinforces the network's image of excellence in global news coverage, and a state highway map is a convenient vehicle for a political message from the governor, image-building photos of the state's tourist attractions, and a cartographic statement about tax dollars being well spent on roads, recreation sites, and forest preserves. An online map that helps voters find their polling places also signifies politicians' delight in gerrymandered boundaries that put incumbents in safe districts. A local map titled "Risk of Rape" can shock and can advocate more diligent police patrols and stricter sentencing, or steer home buyers away from striving, largely safe neighborhoods with an undeserved notoriety. And a cartogram comparing wealth or life expectancy among the world's nations can foster complacent pride or evoke compassionate guilt.

Maps with dual roles are not inherently bad. Indeed, some perfectly correct maps exist primarily to lend an aura of truth, and others exist largely as visual decoration. Decades ago, the impetus for an increased use of news maps was the perception among publishers that a better "packaged," more graphic newspaper could compete effectively with television as well as with rival papers. Their motivation might not have been better reporting, but the conscious decision to use more maps improved their coverage of many news stories in which location was important. Eventually this trend led to the proliferation of news maps online and in other digital media. Maps intended to decorate or impress can educate a public appallingly ignorant about basic place-name geography. Were it not for the map's power as a symbol of geographic knowledge, we would know a great deal less about our neighborhoods, our nation, and the world.

Dual motives are risky, of course. Map authors pursuing aesthetic goals might violate cartographic principles or suppress important but artistically inconvenient information. Maps, like buildings, suffer when the designer puts form ahead of function. Map authors with political or commercial motives might suppress inconvenient information about competing ideologies or sellers as well as knowingly adopt an inappropriate projection or dysfunctional symbols. And expedient map authors who are distracted by a need to decorate can deliver sloppy, misleading maps. The skeptical map viewer will assess the map author's motives and ask how the need to impress might have subverted the need to inform.

Recognizing the map's versatility and its potential to play dual roles should enhance the informed map viewer's healthy skepticism about a map author's expertise or motives. But neither this recognition nor the map's demonstrated ability to distort and mislead should detract from an appreciation of the map's power to explore and explain geographic facts. White lies are an essential element of cartographic language, an abstraction that is enormously useful for analysis and communication. Like verbal language and mathematics, though, cartographic abstraction has costs as well as benefits. If not harnessed by someone who is knowledgeable and with honest intent, the power of maps can get out of control.

Acknowledgments

Many friends, colleagues, and authors have helped hone my interest in maps into the wary enthusiasm I hope to communicate in this book. For the first and second editions I acknowledged a special debt of gratitude

to Darrell Huff, who demonstrated with *How to Lie with Statistics* that a cogent, well-illustrated essay on a seemingly arcane topic could be both insightfully informative and enjoyably readable;

to Syracuse University, for a one-semester sabbatical leave during which much of this book was written;

to Mike Kirchoff, Marcia Harrington, and Ren Vasiliev, for supplementary artwork and for critical comments on the design and execution of my illustrations;

to John Snyder, for information and insights about map projections;

to David Woodward, for continued encouragement; and

to Penny Kaiserlian, for the opportunity to expand and add color.

For this third edition, I add notes of appreciation

to Steven Manson and Ian Muehlenhaus, for supportive comments and useful suggestions;

to Joe Stoll, for a heroic effort to restore images used in the previous edition;

ACKNOWLEDGEMENTS

to John Olson and Darle Balfoort, at Syracuse University's Bird Library, for assistance in locating maps for rescanning;

to Matthew Edney and Mary Pedley, for continued encouragement; and

to Mary Laur, for convincing me that a revision was overdue.

Flattened at the poles and bulging slightly at the equator, the earth is not a perfect sphere. As on the outside of a carousel, centrifugal force is strongest along the equator, where the radius of rotation is greatest. The rotating planet deforms like a highly viscous fluid—like a ball of clay on a potter's wheel. Geodesists have found that the radius from the center of the earth to either pole is shorter by about 1/300 of the radius from the center of the earth to the equator. Although large-scale maps must take this deformation into account, small-scale maps of states, countries, or the entire world can safely and conveniently treat the earth as a sphere.

Latitude and longitude describe positions on the sphere. Longitude is somewhat arbitrary, anchored by international agreement,

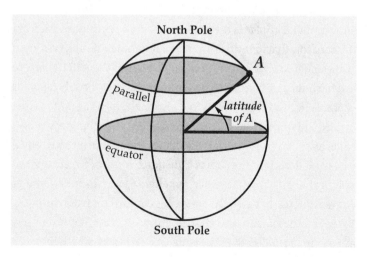

Figure A.1. Spherical earth showing equator, poles, and a parallel and its latitude.

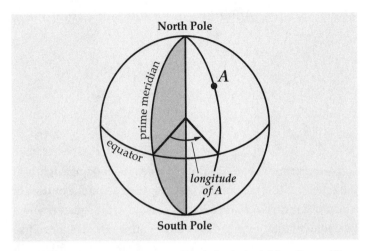

Figure A.2. Spherical earth showing a meridian, its longitude, and the prime meridian.

whereas latitude is a natural coordinate, related to the earth's rotation about its axis. Figure A.1 shows the equator in a plane through the center of the earth and perpendicular to the axis. Point A lies in another plane, parallel to the equator and also perpendicular to the axis; the circle formed where this plane intersects the sphere is a *parallel*. Like all points on the sphere, point A lies in one and only one parallel. *Latitude* is the angle, measured north or south from the equator, that identifies a particular parallel. It ranges from 0° at the equator to 90° at the poles and requires the letter N or S to establish its position north or south of the equator. Chicago, for example, is at 42° N, whereas Sydney, Australia, is at 34° S.

The equator is a *great circle*, the largest circle that occurs on a spherical surface. A great circle divides the sphere into two equal parts and describes the shortest-distance route between any two points along its circumference. A sphere has an infinite number of great circles, but only the equator is equidistant from the poles. Except for the equator, the parallels are *small circles*, defined simply as any circle smaller than the equator. *Meridians*, which intersect the parallels at right angles, are halves of great circles running from pole to pole, as figure A.2 shows. Except at the poles, each point on the sphere has a single meridian and a single parallel.

Longitude, the other spherical coordinate, is the angle that identifies a meridian. Measured east or west as indicated by the letter E or W, it ranges from 0° at the *prime meridian* to 180° at the approximate location of the International Date Line. New York City, for example, is at 74° W, whereas the longitude of Moscow is about 38° E. In 1884, at the International Meridian Conference held in Washington, DC, twenty-two of the twenty-five nations represented endorsed a prime meridian through the Royal Observatory at Greenwich, England. Subsequent international acceptance of the Greenwich meridian ended an era of cartographic isolation marked by prime meridians through Cádiz, Christiania, Copenhagen, Ferro, Lisbon, Naples, Paris, Pulkowa, Rio de Janeiro, and Stockholm, among others. Historians and other users of old maps must be particularly conscious of prime meridians and longitude.

Selected Readings for Further Exploration

Readers interested in further information about maps and their uses might find this varied list of sources helpful.

Elements of the Map

Bunge, William. *Theoretical Geography*. Lund Studies in Geography, series C, General and Mathematical Geography, no. 1. Lund, Sweden: C. W. K. Gleerup, 1962. Especially chapter 2, "Metacartography."

Dodge, Martin, Rob Kitchin, and Chris Perkins, eds. *The Map Reader: Theories of Mapping Practice and Cartographic Representation*. Chichester, West Sussex, UK: Wiley-Blackwell, 2015.

Goodchild, Michael F. "Scale." In *Cartography in the Twentieth Century* (Volume Six of the *History of Cartography*), edited by Mark Monmonier (hereafter referred to as *HC6*), 1383–89. Chicago: University of Chicago Press, 2015.

Greenhood, David. *Mapping*. Chicago: University of Chicago Press, 1964.

Keates, J. S. *Understanding Maps*. New York: John Wiley, 1982.

MacEachren, Alan M. *How Maps Work: Representation, Visualization, and Design*. New York: Guilford Press, 2004.

———. *Some Truth with Maps: A Primer on Symbolization and Design*. Washington, DC: Association of American Geographers, 1994.

Monmonier, Mark. *Mapping It Out: Expository Cartography for the Humanities and Social Sciences*. Chicago: University of Chicago Press, 1993.

Robinson, Arthur H., and Barbara Bartz Petchenik. *The Nature of Maps: Essays toward Understanding Maps and Mapping*. Chicago: University of Chicago Press, 1976.

Slocum, Terry A., Robert B. McMaster, Fritz C. Kessler, and Hugh H. Howard. *Thematic Cartography and Geovisualization*. 3rd ed. Upper Saddle River, NJ: Pearson Prentice Hall, 2014.

Snyder, John P. *Map Projections—A Working Manual*. US Geological Survey Professional Paper 1395. Washington, DC: US Government Printing Office, 1987.

Tyner, Judith A. *Principles of Map Design*. New York: Guilford Press, 2010.

History of Cartography

Blakemore, M. J., and J. B. Harley. *Concepts in the History of Cartography: A Review and Perspective.* Cartographica Monographs, no. 26. Toronto: University of Toronto Press, 1980.

Davies, John, and Alexander J. Kent. *The Red Atlas: How the Soviet Union Secretly Mapped the World.* Chicago: University of Chicago Press, 2017.

Edney, Matthew H. "Histories of Cartography." *HC6,* 607–14.

Harley, J. B., and David Woodward, eds. *The History of Cartography.* 6 vols. Chicago: University of Chicago Press, 1987–.

Monmonier, Mark. *Maps with the News: The Development of American Journalistic Cartography.* Chicago: University of Chicago Press, 1989.

Rankin, William. *After the Map: Cartography, Navigation, and the Transformation of Territory in the Twentieth Century.* Chicago: University of Chicago Press, 2016.

Robinson, Arthur H. *Early Thematic Mapping in the History of Cartography.* Chicago: University of Chicago Press, 1982.

Snyder, John P. *Flattening the Earth: Two Thousand Years of Map Projections.* Chicago: University of Chicago Press, 1993.

Thrower, Norman J. W. *Maps and Civilization: Cartography in Culture and Society.* 4th ed. Chicago: University of Chicago Press, 2017.

Woodward, David, ed. *Art and Cartography: Six Historical Essays.* Chicago: University of Chicago Press, 1987.

Map Use and Map Appreciation

Kent, Alexander J., and Peter Vujakovic. *The Routledge Handbook of Mapping and Cartography.* London: Routledge, 2018.

Kimerling, A. Jon, Aileen R. Buckley, Phillip C. Muehrcke, and Juliana O. Muehrcke. *Map Use: Reading, Analysis, and Interpretation.* 8th ed. Redlands, CA: Esri Press, 2016.

Kjellström, Björn, and Carina Kjellström Elgin. *Be Expert with Map and Compass: The Complete Orienteering Handbook.* 3rd ed. Hoboken, NJ: John Wiley & Sons, 2010.

Maling, D. H. *Measurements from Maps: Principles and Methods of Cartometry.* Oxford: Pergamon Press, 1989.

Monmonier, Mark, and George A. Schnell. *Map Appreciation.* Englewood Cliffs, NJ: Prentice-Hall, 1988.

Map Generalization

Eckert, Max. "On the Nature of Maps and Map Logic." Translated by W. Joerg. *Bulletin of the American Geographical Society* 40 (1908): 344–51.

Jenks, George F. "Lines, Computers, and Human Frailties." *Annals of the Association of American Geographers* 71 (1981): 1–10.

McMaster, Robert B., and K. Stuart Shea. *Generalization in Digital Cartography.* Washington, DC: Association of American Geographers, 1992.

Muller, Jean-Claude. "Generalization of Spatial Data Bases." In *Geographical Information Systems: Principles and Applications*, edited by David Maguire, Michael Goodchild, and David Rhind, 457–75. London: Longman, 1991.

Wright, John K. "Map Makers Are Human: Comments on the Subjective in Maps." *Geographical Review* 32 (1942): 527–44.

Color

Brewer, Cynthia A. *Designing Better Maps: A Guide for GIS Users.* 2nd ed. Redlands, CA: Esri Press, 2016.

English-Zemke, Patricia. "Using Color in Online Marketing Tools." *IEEE Transactions on Professional Communication* 31 (1988): 70–74.

Falk, David S., Dieter R. Brill, and David G. Stork. *Seeing the Light: Optics in Nature, Photography, Color, Vision, and Holography.* New York: Harper and Row, 1986. Especially pp. 238–86.

Hardin, C. L. *Color for Philosophers: Unweaving the Rainbow.* Indianapolis, IN: Hackett Publishing Co., 1988.

Hoadley, Ellen D. "Investigating the Effects of Color." *Communications of the ACM* 33 (1990): 120–25, 139.

Olson, Judy M. "Color and the Computer in Cartography." In *Color and the Computer*, edited by H. John Durrett, 205–19. Boston: Academic Press, 1987.

Development Maps and Geographic Information Systems

Aberley, Doug, ed. *Boundaries of Home: Mapping for Local Empowerment.* Philadelphia, PA: New Society Publishers, 1993.

Armstrong, Marc P., Paul J. Densham, Panagiotis Lolonis, and Gerard Rushton. "Cartographic Displays to Support Locational Decision Making." *Cartography and Geographic Information Systems* 19 (1992): 154–64.

Goodchild, Michael F., Bradley O. Parks, and Louis T. Steyaert. *Environmental Modeling with GIS.* New York: Oxford University Press, 1993.

Pickles, John, ed. *Ground Truth: The Social Implications of Geographic Information Systems.* New York: Guilford Press, 1995.

Maps for Advertising and Political Propaganda

Akerman, James R., ed. *Cartographies of Travel and Navigation.* Chicago: University of Chicago Press, 2006.

Bassett, Thomas J. "Cartography and Empire Building in Nineteenth-Century West Africa." *Geographical Review* 84 (1994): 316–35.

Boggs, S. W. "Cartohypnosis." *Scientific Monthly* 64 (1947): 469–76.

Davis, Bruce. "Maps on Postage Stamps as Propaganda." *Cartographic Journal* 22 (1985): 125–30.

Fleming, Douglas K. "Cartographic Strategies for Airline Advertising." *Geographical Review* 74 (1984): 76–93.

Gilmartin, Patricia. "The Design of Journalistic Maps: Purposes, Parameters, and Prospects." *Cartographica* 22 (Winter 1985): 1–18.

Herb, Guntram. "Geopolitics and Cartography," *HC6*, 539–48.

McDermott, Paul D. "Advertising, Maps as." *HC6*, 18–22.

Monmonier, Mark. *Drawing the Line: Tales of Maps and Cartocontroversy*. New York: Henry Holt and Co., 1995.

———. "The Rise of the National Atlas." *Cartographica* 31 (Spring 1994): 1–15.

Quam, Louis O. "The Use of Maps in Propaganda." *Journal of Geography* 42 (1943): 21–32.

Speier, Hans. "Magic Geography." *Social Research* 8 (1941): 310–30.

Tyner, Judith A. "Persuasive Cartography." *HC6*, 1087–95.

Maps, Defense, and Disinformation

Clarke, Keith C. "Maps and Mapping Technologies of the Persian Gulf War." *Cartography and Geographic Information Systems* 19 (1992): 80–87.

Demko, G. J., and W. Hezlep. "USSR: Mapping the Blank Spots." *Focus* 39 (Spring 1989): 20–21.

Paglen, Trevor. *Blank Spots on the Map: The Dark Geography of the Pentagon's Secret World*. New York: Dutton, 2010.

Stommel, Henry. *Lost Islands: The Story of Islands That Have Vanished from Nautical Charts*. Vancouver: University of British Columbia Press, 1984. Especially chapter 4, "The Fake Island of Captain Benjamin Morrell."

National Mapping and the Cartographic Subculture

Edney, Matthew H. "Cartography without 'Progress': Reinterpreting the Nature and Historical Development of Mapmaking." *Cartographica* 30 (Summer/Autumn 1993): 54–68.

———. "Politics, Science, and Government Mapping Policy in the United States, 1800–1925." *American Cartographer* 13 (1986): 295–306.

Harley, J. B. *Maps and the Columbian Encounter*. Milwaukee: Golda Meir Library, University of Wisconsin, 1990.

Larsgaard, Mary Lynette. *Topographic Mapping of the Americas, Australia, and New Zealand*. Littleton, CO: Libraries Unlimited, 1984.

Lewis, G. Malcolm, "Metrics, Geometries, Signs, and Language: Sources of

Cartographic Miscommunication between Native and Euro-American Cultures in North America." *Cartographica* 30 (Spring 1993): 98–106.

McHaffie, P. H. "The Public Cartographic Labor Process in the United States: Rationalization Then and Now." *Cartographica* 30 (Spring 1993): 55–60.

Rundstrom, Robert A. "The Role of Ethics, Mapping, and the Meaning of Place in Relations between Indians and Whites in the United States." *Cartographica* 30 (Spring 1993): 21–28.

Schulten, Susan. *Mapping the Nation: History and Cartography in Nineteenth-Century America*. Chicago: University of Chicago Press, 2012.

Southard, R. B. "The Development of U.S. National Mapping Policy." *American Cartographer* 10 (1983): 5–15.

Thompson, Morris M. *Maps for America: Cartographic Products of the U.S. Geological Survey and Others*. Reston, VA: US Geological Survey, 1979.

Woodward, David. "Map Design and the National Consciousness: Typography and the Look of Topographic Maps." *Technical Papers of the American Congress on Surveying and Mapping* (Spring 1992): 339–47.

Data Maps

Fisher, Howard T. *Mapping Information: The Graphic Display of Quantitative Information*. Cambridge, MA: Abt Books, 1982.

Kitchen, Rob. *The Data Revolution: Big Data, Open Data, Data Infrastructures and Their Consequences*. Los Angeles: SAGE Publications, 2014.

Monmonier, Mark. "The Hopeless Pursuit of Purification in Cartographic Communication: A Comparison of Graphic-arts and Perceptual Distortions of Graytone Symbols." *Cartographica* 17 (1980): 24–39.

Tobler, W. R. "Choropleth Maps without Class Intervals?" *Geographical Analysis* 5 (1973): 262–65.

Tufte, Edward R. *Envisioning Information*. Cheshire, CT: Graphics Press, 1990.

———. *The Visual Display of Quantitative Information*. Cheshire, CT: Graphics Press, 1983.

Image Maps

Biesecker, Barbara. "No Time of Mourning: The Rhetorical Production of the Melancholic Citizen-Subject in the War on Terror." *Philosophy and Rhetoric* 40 (2007): 147–69. (The quotations from Colin Powell's speech to the UN Security Council are on pp. 159–61.)

Buchroithner, Manfred F. "Remote Sensing: Satellite Systems for Cartographic Applications." *HC6*, 1282–88.

Collier, Peter. "Photogrammetric Mapping: Orthophotography and Orthophoto Mapping." *HC6*, 1141–46.

Dozier, Jeff. "Remote Sensing: Earth Observation and the Emergence of Remote Sensing." *HC6*, 1273–82.

Morris, Errol. "Photography as a Weapon." Opinion section, *New York Times*, August 11, 2008.

Xie, Yichun, Zongyao Sha, and Mei Yu. "Remote Sensing Imagery in Vegetation Mapping: A Review." *Journal of Plant Ecology* 1 (2008): 9–23.

Prohibitive Cartography

Chekovich, Alex. "Land Use Mapping." *HC6*, 751–54.

Ehrenberg, Ralph E. "Aeronautical Chart." *HC6*, 22–30.

Monmonier, Mark. *Cartographies of Danger: Mapping Hazards in America*. Chicago: University of Chicago Press, 1997.

———. *No Dig, No Fly, No Go: How Maps Restrict and Control*. Chicago: University of Chicago Press, 2010.

———. *Spying with Maps: Surveillance Technologies and the Future of Privacy*. Chicago: University of Chicago Press, 2002.

Fast Maps

Battersby, Sarah E., Michael P. Finn, E. Lynn Usery, and Kristina H. Yamamoto. "Implications of the Web Mercator and Its Use in Online Mapping." *Cartographica* 49.2 (2014): 85–101.

Fish, Caroline S., and Kirby Calvert. "Analysis of Interactive Solar Energy Web Maps for Urban Energy Sustainability." *Cartographic Perspectives*, no. 85 (2016): 5–22.

Goodchild, Michael F., and Linna Li. "Assuring the Quality of Volunteered Geographic Information." *Spatial Statistics* 1 (2012): 110–20.

Haughney, Christine. "After Pinpointing Gun Owners, Paper Is a Target." *New York Times*, January 6, 2013.

Ingram, Matthew. "Newspapers and Guns: If Data Is Available, Should It Always Be Published?" Gigaom.com, December 27, 2012, https://gigaom.com/2012/12/27/newspapers-and-guns-if-data-is-available-should-it-always-be-published/.

Muehlenhaus, Ian. *Web Cartography: Map Design for Interactive and Mobile Devices*. Boca Raton, FL: CRC Press, 2014.

Ortag, Felix, Manuela Schmidt, and Georg F. Gartner. "Web-based Wayfinding," *HC6*, 1739-42.

Perkins, Chris. "Plotting Practices and Politics: (Im) Mutable Narratives in OpenStreetMap." *Transactions of the Institute of British Geographers* 39 (2014): 304–17.

Peterson, Michael P. *Mapping in the Cloud*. New York: Guilford Press, 2014.

Sources of Illustrations

3.4 (left): US Geological Survey, 1973, Northumberland, PA, 7.5-minute quadrangle map.

3.4 (right): US Geological Survey, 1969, Harrisburg, PA, 1:250,000 series topographic map.

3.8 (left): New York State Department of Transportation, 1980, Rochester East (north part), 1:9,600 urban area map.

4.1: Compiled from maps in Brian J. Hudson, "Putting Granada on the Map," *Area* 17, no. 3 (1985): 233–35.

8.1: Christopher Saxton, *Atlas of England and Wales* (1579), and Maurice Bouguereau, *Le théâtre françoys* (1594), atlases in the collection of the US Library of Congress.

8.2: *Area Handbook for Pakistan* (Washington, DC: US Government Printing Office, 1965), p. xii.

8.3: American University Digital Research Archive.

8.7: Courtesy of John P. Snyder (according to Snyder, the projection was devised about 1945 and first shown to cartographers in 1987; Snyder has not published it).

8.8: *Facts in Review* 1, no. 17 (December 8, 1939): 1.

8.9: *Facts in Review* 3, no. 16 (May 5, 1941): 250.

8.10: *Facts in Review* 2, no. 5 (February 5, 1940): 33.

8.11: *Facts in Review* 2, no. 28 (July 8, 1940): 294.

8.12: *Facts in Review* 1, no. 16 (November 30, 1939): 3.

8.13: *Facts in Review* 3, no. 13 (April 10, 1941): 182.

8.15: Associated Press Wirephoto map, used on the front page of the *Syracuse Herald-Journal* for July 6, 1950; reprinted with the permission of the Associated Press.

8.17: David M. Smith, *Where the Grass Is Greener* (Baltimore, MD: Johns Hopkins University Press, 1982), fig. 1.6, © David M. Smith, 1979; reprinted by permission.

9.1: Compiled from facsimile maps in "Soviet Cartographic Falsification," *Military Engineer* 62, no. 410 (November–December 1970): 389–91, figs. 2a–e.

9.2: Compiled from facsimile maps in "Soviet Cartographic Falsification,"

Military Engineer 62, no. 410 (November–December 1970): 389–91, figs. 1a–c.

9.3: New York State Department of Transportation, 1978, Rome (north), 1:9,600 urban-area map.

9.4 (map): US Geological Survey, 1985, Blue Ridge Summit, MD-PA, 7.5-minute quadrangle map.

9.4 (quotation): Edward C. Papenfuse et al., *Maryland: A New Guide to the Old Line State* (Baltimore, MD: Johns Hopkins University Press, 1976), p. 64.

9.5 (left): US Army Map Service, 1946, Tonawanda West, NY, 7.5-minute quadrangle map.

9.5 (right): US Geological Survey, 1980, Tonawanda West, NY, 7.5-minute quadrangle map.

10.1: US Geological Survey, 1902, Everett, PA, 15-minute quadrangle map.

10.4 (left): US Geological Survey, 1980, Bunker Hill, IN, 7.5-minute quadrangle map (photorevised).

10.4 (right): US Geological Survey, 1994, Bunker Hill, IN, 7.5-minute quadrangle map.

10.5: US Geological Survey, 1982, North Haven East, ME, 7.5-minute quadrangle map (provisional).

10.6: US Geological Survey, 1986, Scranton, PA, 1:100,000 planimetric map.

10.7: US Geological Survey, 1950, 1957, 1965, and 1969 editions, Harrisburg, PA, 1:250,000 topographic map.

10.8 (above): US Geological Survey, 1958, Cajon Mesa, UT-CO, 15-minute quadrangle map.

10.8 (below): US Geological Survey, 1985, Wickiup Canyon, UT-CO, 7.5-minute quadrangle map.

14.2: OpenStreetMap.org.

Plate 1: Metrorail System Map, courtesy of Washington Metropolitan Area Transit Authority.

Plate 12 (left): US Geological Survey, 2000, Bath, ME, 7.5-minute quadrangle map (photorevised).

Plate 12 (right): US Geological Survey, 1999, Bath-SW, ME, digital orthophoto quadrangle map.

Plate 13: NASA Earth Observatory image created by Jesse Allen, using data provided courtesy of NASA/GSFC/METI/ERSDAC/JAROS, and US/Japan ASTER Science Team.

Plates 14 and 15: US Department of State.

Plate 16: Matthew Ingram, "Newspapers and Guns: If Data Is Available, Should It Always Be Published?" Gigaom.com, December 27, 2012.

Plate 17: Google Maps (google.com/maps).

Index

The letter *f* following a page number denotes a figure.

61, 62f, plate 2; distortions caused by black-and-white printing, 67–68; distortions on color monitors, 63–64, 71, plate 4; effectiveness as a map symbol, 69–70; HVS color-space diagram, 61–62, 62f; hypsometric tints, 70, plate 10; image maps, 181; inadvertent camouflage, 71; printing distortions, 64–65, plate 4; problem in using hue to portray differences in measures, 65–66, 66f; roles of hue and value, 62; simultaneous contrast and, 70, plate 11; spectral-scale use, 66–67, plate 7; subtractive primary colors, 63, plate 3; symbols and, 24–26, 69–70; technological change enabling the misuse of, 60; televisions and, 61, 64

color wheel, 61, 62f

commercial mapmakers, 20, 38, 52

community planning. *See* development maps

computer mapping systems. *See* graphics software

concept diagrams for developments, 92–93, 92f

conformal projections, 14–15, 199

conic projections, 12

content generalization: bias in generalized land-cover maps, 41–44, 43f, 44f; bias potential of software-generated maps, 40–42, 41f; classification element, 37, 38–39; example of how a template can mislead, 38, 39f; experimentation using graphics software, 37, 40, 41, 195–96; intuition and ethics in, 44; selection element, 37; on soils maps, 38–40; and standardized symbols, 37–38

contour lines, 22–23, 22f, 136, 137f, 147

copyright infringement, 52

correlation: coefficient of, 170; geographic, 170–71, 171f; grid lines manipulation and, 165–68, 166f, 167f; statistical, 168–70, 169f

culture and maps. *See* mapping culture; public culture

Cuomo, Mario, 133

curvilinear correlation, 170

cylindrical projections, 9f, 10–13, 11f, 12f, 201

data maps: areal aggregation over time and, 174; areal aggregation's effect on geographic patterns, 156–57, 156f; areal aggregation's reliance on averages, 158; areal aggregation when mapping patterns of rates and ratios, 154f, 155–56, 155f; areally aggregated data and its value, 159; areal units' ability to hide spatial detail, 157, 158f, 168; areal units selection and, 153–54; cautions for using aggregated data, 159; choosing indexes, rates, and point symbols to communicate a position, 175; choosing meaningful breaks, 163–65, 164f; classifications use, 159–63, 160–62f; correlation depictions and grid lines manipulation, 165–68, 166f, 167f; demographic data, 176–77, 177f; disease maps and associated cautions, 172–74, 173f; ease of misleading with, 153; ecological fallacy vulnerability, 171–72; geographic correlation, 170–71, 171f; number line, 163–66, 164f; one-map solutions and associated dangers, 174–75; outliers treatment, 164–65; seasonal adjustments, 176; spatial patterns selection, 175–76, 176f; statistical correlations and their depictions, 168–70, 169f

demographic base map, 18–19, 112, 174, 177

Department of State, US, 188

derivative maps, 46

development maps: challenging a property assessment using, 97–99, 98–100f; community planning boards and, 86–88, 87f; concept diagrams, 92–93, 92f; environmental-regulations

107, 108f; legitimacy conferred by maps, 102, 104; molding of a map's message, 101–2; naming use on, 121–22, 122f; Nazi propaganda, 112–18; opposition to equal-area projections, 110–12, 111f; reflecting political bias in a map, 42–44, 43f, 44f; size distortion use on, 107–12, 109f, 110f; symbols of power and nationhood, 102–4; territory disputes and, 104–7, 105f; titles and, 176. *See also* advertising (ad) maps

property assessments: challenging using maps, 97–99, 98–100f; process in municipalities, 95–97

property maps, 189–90

provisional maps, 145, 146–47, 146f

public culture, 136. *See also* mapping culture

quadrangles, 135, 136, 144–45, 147

quantile schemes, 162

quartiles, 161

quintiles, 162

radiometers, 181–82

railroad symbols, 8, 33, 73–75, 136–40, 137f, 139f, 149

raster data, 179–80, plate 12

ratio scales, 5, 9, 10–11, 10f

real-estate development. *See* development maps

redacting, 185–86, 185f

relief displacement, 35

restrictive maps. *See* prohibitive cartography

RGB color cube, 64

rhumb lines, 16, 107

Richard, Mount (prank), 53–54

road maps: cartographic carelessness on, 46, 50–52, 51f; commercial, 39f; creation of, 151; generalization and, 32; symbols on, 20

Russia, 108, 127, 127f

Salmi, Russia, 127, 127f

saturation, 61, 62f, plate 2

Saxton, Christopher, 103

scale on maps: detail capacity of ratio scales, 5, 9; distortion and the standard line, 11–12; graphic, 7–8; large- versus small-scale maps, 6–7; as a means of comparison, 5–6; metric units and, 7, 181; types of, 5, 6f; verbal statements of and their use and misuse, 7; zoom slider and, 8, 181

scatterplots, 165–67, 167f, 169, 169f, 170–71, 171f

seasonal adjustments of data, 176

secant conic projections, 12

secant cylindrical projections, 11

security. *See* cartographic security

segmentation of an area, 32

selection: of content, 37–38, 40, 101, 196; of features, 20, 27–33, 137

shape: conformality and, 199; development maps' use of, 85; distortion in projections, 13, 14–15, 17, 110, 111f, 112; errors in, 50, 73; humorous distortions' use of, 81; of symbols, 20–21, 21f, 22, 38

simplification of features, 29

simultaneous contrast, 70, plate 11

single-place focus maps: geometric distortion and place-name dropping on, 77–79, 78f; and humorous distortions, 80–81; pictorial icons on, 79–80, 80f; purpose of an ad map, 77

sinusoidal projections, 13–14

size symbols, 21, 21f, 22

slider, zoom, 8, 181

slippy maps, 199

slope on topographic maps, 22–23, 22f, 88, 136, 140

small-scale maps, 6–7

smoothing of features, 29

Snow, John, 173

Snyder, John, 111

social activists' use of maps, 122

software developers, 1, 2, 42. *See also* fast maps; graphics software

soils maps: aggregation on, 32; color on, 65; content-generalization cautions, 38–40; environmental assessments and, 90, 91f; regional planning and, 85; size threshold on, 32; use cautions, 90, 91f. *See also* geologic maps